THE AUDIOPHILE LOUDSPEAKER

Anyone Can Build

by Gene Healy

Boston Post Publishing Company Madison, Connecticut

First Edition
First Printing

Boston Post Publishing Company
P.O. BOX 1175 MADISON, CONNECTICUT 06443

Library of Congress Catalog Number: 95-77721

ISBN 0-9647777-0-3

Editorial Director Abigail Pattee
Editors Christine Meehan, Robert Fort, J.R. Cochran
Book design Audrey Bennett
Cover design Audrey Bennett
Photography Michael Marsland
Diagrams Rod Recor
Technical assistance Don Butler, Vincent Oneppo

This book is dedicated to my friends

This book is separated into segments. Each segment describes particular components and how they are connected and interact. It is important to read the entire text before you attempt assembly so that you will have a comprehensive view of all aspects of this simple project.

For ease of assembly during construction, you may want to refer to the appropriate segment listed in the table of contents.

Contents

Introduction

Building a high-end loudspeaker system is a rewarding pastime that does not have to cost a lot of money. As a matter of fact, constructing your own high-quality speaker system can cost considerably less and sound better than one purchased complete. This manual will show you the easy way to build a great home stereo loudspeaker system that will give you years of listening pleasure.

We would like you to know that you don't need to be an engineer to build a superlative loudspeaker system. Understanding loudspeaker design can be easy or complicated. The complex type of system instruction assumes prior knowledge of technical definitions and terms or the education and background to understand the jargon. *The Audiophile Loudspeaker Anyone Can Build* contains no hard-to-understand, technical terminology or schematics. Instead, it outlines all aspects of a great loudspeaker project in plain, easy-to-understand language. This book presumes the reader knows nothing about building loudspeakers and sticks to the "how to build it" aspects of a first-rate speaker system with easy-to-read text, photographs and diagrams.

It was our goal to make this unique manual as simple and understandable as possible.

All the parts necessary to complete the system described in the book, with the exception of the loudspeaker enclosures (boxes), are pre-assembled. Pre-assembled parts can be purchased over the counter at your local electronics store, hardware store or through the mail.

Companies that sell loudspeaker components and parts are listed in the back of the manual.

I

The System

The stereo loudspeaker system described in this manual is called a two-way system. This means each of the two loudspeaker enclosures contains two drivers or loudspeakers. The drivers are a 6-1/2" **woofer** and a smaller loudspeaker called a **tweeter.** It is difficult to beat the accuracy of a 6-1/2" two-way loudspeaker system. Recording and broadcast studios often use them as monitors.

Two 6-1/2" woofers (front and side views).

Two 1" soft-dome tweeters (front and back).

A woofer is a loudspeaker designed to reproduce low-frequency sounds. Bass guitars and kettle drums produce low frequency sounds. A tweeter is a loudspeaker designed to reproduce high frequency sounds. Violins and flutes produce high frequency sounds.

Sound frequency vibrations are measured in hertz or Hz. People can hear sounds from about 20 Hz to 20,000 Hz. Twenty vibrations per second are very low sounds. Twenty-thousand vibrations per second are very high sounds.

The most relevant loudspeaker measurement is its frequency range. The quality of a loudspeaker is determined by how smoothly it goes from low Hz to higher Hz.

The measurement of loudspeaker frequency range often appears as a line across a graph. Generally speaking, the straighter the line (the fewer big dips and peaks) the better the speaker. This information is useful when selecting loudspeakers.

In addition to a woofer and a tweeter, each enclosure will contain three simple electrical devices and two inert (non-electrical) acoustic components. A crossover, L-Pad and terminal cup are the electrical parts. A port tube and some synthetic damping material, which will affect the acoustical nature of the loudspeaker enclosure (box), are the inert or non-electrical parts.

The loudspeaker system described in this manual is versatile. The enclosures are designed to accept any number of fine loudspeaker components available over the counter or through the mail. You will be pleased to know that the quality of even some of the lower priced loudspeakers (drivers) and components available today is so high that chances are you will be well pleased with almost any combination you choose.

Note: *If you are not handy or don't have the tools, please don't be intimidated by the construction of the loudspeaker enclosures (boxes). With the plans and directions in this manual, any carpenter, cabinetmaker or experienced friend can make these simple boxes for you to your specifications. If you don't know anyone who can help, look in your phone directory.* ***Loudspeakers and other electrical components used in this system will be 8 ohms. An ohm is a measurement of electrical resistance. Eight-ohm loudspeakers are easy for an amplifier to power or drive.***

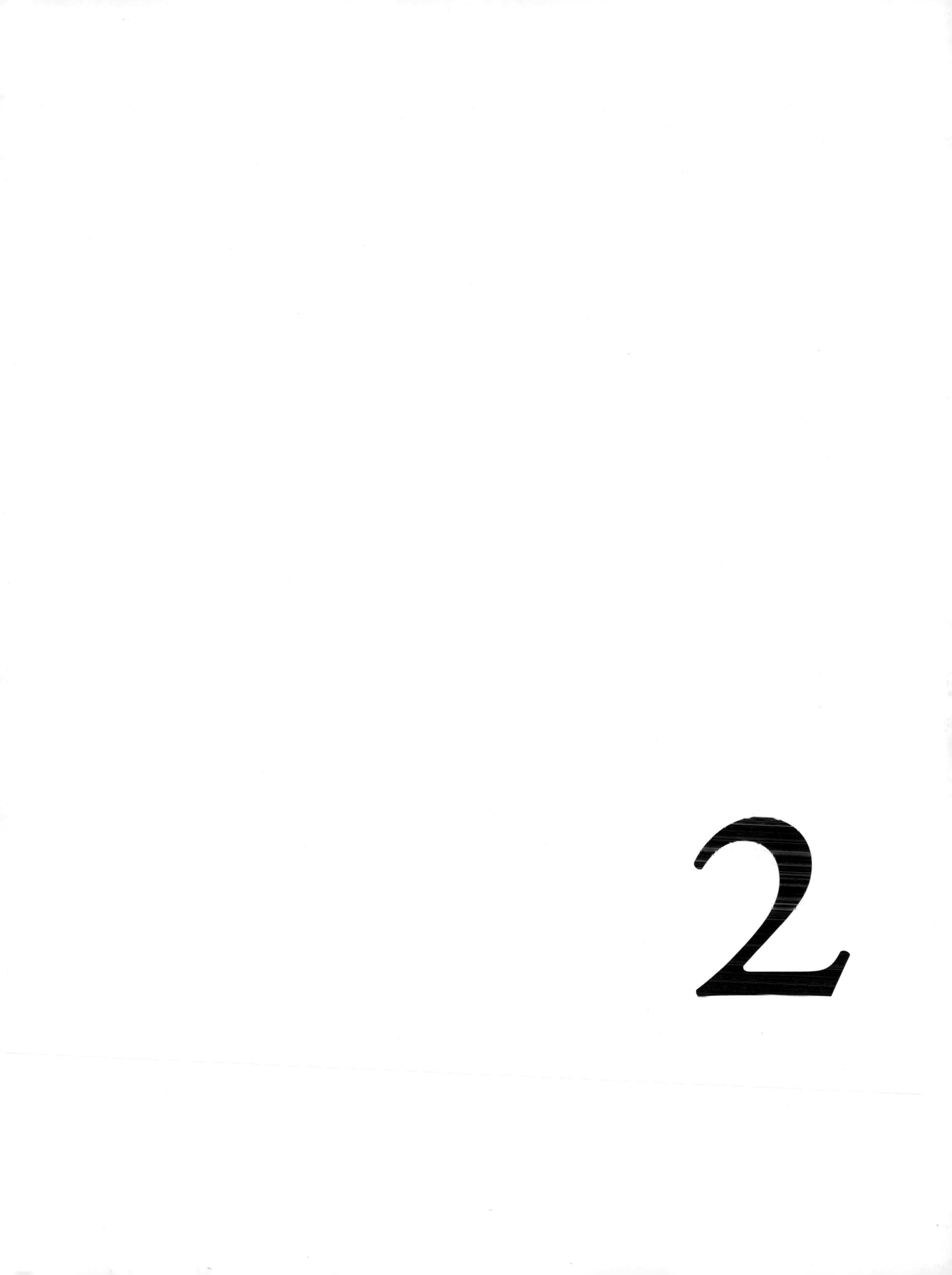

Enclosure shown with 1" front baffle and 3/4" rear panel.

The Enclosures

Each loudspeaker enclosure (box) has an internal volume of approximately .92 cubic feet or 26 liters.

The external (outside) dimensions of the enclosures are: 9" wide x 19-1/2" high x 13-3/4" deep. The internal (inside) dimensions of the enclosures are: 7-1/2" wide x 18" high x 12" deep.

Butt joint construction (detail).

Butt joint loudspeaker enclosure with brace installed.

Note: The internal or inside dimensions are more important than the outside dimensions. Keep the inside dimensions as close to specifications as possible. You have a 10% plus or minus margin of error to play with. In other words, if your box is 10% smaller than specified it will be okay. If it's 10% larger, that will be okay, too.

A rule of thumb in loudspeaker construction is that you don't want the box you build to resonate or vibrate. It should be solid and braced.

The enclosed plan illustrates butt joint construction with corner blocks. However, we want you to know that there are a lot of ways to build a box and it makes no difference what kind of joints you employ or how you build it. It can be mitered, splined, whatever. The most important aspect of the loudspeaker enclosure construction is that it be solid and strong.

Loudspeaker enclosures can be made of plywood, particle board or a product called medium-density fiberboard (MDF). Of the three choices, MDF is considered best for loudspeaker construction. Particle board is second best and plywood is the third choice. MDF and particle board are often used because these materials do not contain voids in plies which might rattle. Plywood is fine if you use a high grade and check it for voids and imperfections in the plies. All of the above materials can be purchased at lumber yards and come in a variety of attractively veneered surfaces.

Generally speaking, the thicker the building material the better. We recommend a minimum thickness of 3/4″ MDF, particle board or plywood for the sides, tops, bottom and back of the enclosures. The front baffle which holds the loudspeakers will be better if it is one inch thick.

Note: *We recommend that you use screws and wood glue. For ease of construction, it is best to pre-drill all holes for screws. Screws act as clamps while the glue sets.*

A repeated word of advice: The enclosed plans are not complicated. However, if you don't have the tools or experience, we recommend that you get out your phone directory and have a carpenter, cabinetmaker or experienced friend build these boxes for you. The end result may be better. The final assembly will be up to you.

3

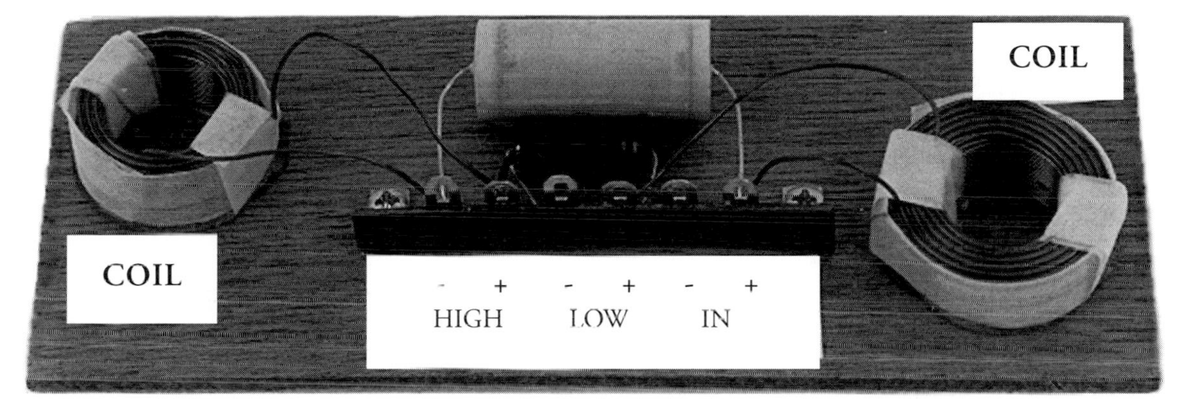

A two-way 12dB crossover.

Crossovers

A crossover network is an electrical filter that separates low frequencies from high frequencies and delivers the separated sounds or frequencies to the woofer and the tweeter. A crossover is attached to the inside of the speaker enclosure as far away from the speaker magnets as possible.

Crossover networks consist of two main elements: coils and capacitors. Coils, also known as inductors, block high sound frequencies and capacitors block low sound frequencies. We will be using 2 two-way, 8 ohm crossovers for this project—one for each loudspeaker.

The woofer or low-pass section of a two-way crossover is designed to deliver low frequency signals to the woofer. Most out-of-the-box commercial two-way crossovers cut low frequency signals to the woofer between 3000 Hz and 5000 Hz. The tweeter or high-pass section of a crossover takes over where the woofer leaves off and sends high frequency sound signals to the tweeter.

Crossover networks can be custom-designed and built for a particular loudspeaker and enclosure volume or purchased pre-assembled and ready to go in a variety of configurations. Pre-assembled crossovers can be purchased over the counter at electronics stores or from catalogues.

For this project, we strongly suggest that you start out with a commercially sold, pre-assembled two-way crossover. It's a good way to familiarize yourself with crossovers. Later on, you may wish to change to a custom-designed crossover.

Computer-generated, custom-designed crossovers often require a knowledge of electrical schematics and the ability to assemble the crossover. Building a custom crossover requires experience. If you are familiar with loudspeaker crossovers and know how to read schematic drawings and assemble electrical components, a custom crossover may be for you. Sometimes companies that offer custom designs will assemble the crossover for you, but often prefer to just send the parts and schematics

Two-way crossover installed on back panel of enclosure between terminal cup and L-Pad..

and leave the assembly to you. If you're just starting out, you will be better off beginning with a pre-assembled crossover.

A commercial pre-assembled crossover generally comes in what is called a 6dB per octave roll-off, or a 12dB per octave roll-off. This is known as the slope of the crossover. A 12dB crossover has a steeper slope than a 6dB crossover. That means that the sounds coming out of a loudspeaker are more sharply delineated at the crossover point with a 12dB crossover. Sometimes crossovers have a 6dB woofer section and a 12dB tweeter section.

Important: *Before purchasing a tweeter and a crossover, it is important to check the tweeter's frequency response and at what point your crossover shifts from low frequencies to high frequencies. In a crossover, this frequency point is referred to as roll-off. Frequency responses for tweeters and roll-off points for two-way crossovers will be listed in catalogues or included with components. Frequency responses of tweeters and roll-off points of crossovers have to be matched correctly. This should be done to make sure you don't damage your tweeter with frequencies that are too low for it to handle after it has been connected to the crossover.*

Matching a tweeter to a commercially sold (out-of-the-box) crossover is simple. All you have to do is make sure the tweeter (high) section of the crossover rolls off roughly two times the resonance frequency of the tweeter when using a 12dB crossover. A 6dB crossover should be used only with tweeters that have a low resonance frequency and should be crossed over between 4000 Hz and 5000 Hz.

Important: *Resonance frequency means the low end of the tweeter's frequency response. If you will be using your loudspeakers at high volume levels, you will be better off using 12dB commercial crossovers or crossovers that have a 12dB tweeter section. A 12dB crossover provides a sharper frequency shift than a 6dB crossover and blocks low frequencies from reaching the tweeter more effectively.*

Example: *A tweeter with a frequency response of 2000 Hz to 20,000 Hz can safely be crossed over at 4000 Hz with a 12dB crossover. When you browse through catalogues you will note that frequency responses of tweeters and crossover points (roll-off) for different crossovers are listed. Improperly crossed-over tweeters are susceptible to damage.*

Note: *6dB crossovers should be used with low resonance tweeters that are crossed over between 4000 Hz and 5000 Hz–5000 Hz being the safer choice.*

Matching a woofer to a crossover is easy. Just make sure that the high end of the woofer's frequency response goes higher than or at least matches the low section (woofer section) of your two-way crossover.

Example: *A 6-1/2" woofer listed with a frequency response of 60 Hz to 4500 Hz can be crossed over effectively anywhere up to 4500 Hz but not at 5000 Hz. It probably would be better to cross the woofer over at 4000 Hz. If you cross the woofer over at 4000 Hz, you won't be stretching the capability of the speaker to reproduce sound accurately.*

Note: *Most 6 1/2" woofers will perform best in a two-way system if the crossover point is between 3000 Hz and 5000 Hz.*

A two-way crossover has six connectors. Two are connected to the woofer (low section), two are connected to the tweeter (high section) and two are connected to the terminal on the back of the loudspeaker enclosure (IN). One of the two connection points for each component (woofer, tweeter and terminal) is positive (+) and one is negative or common (-). A red color mark on a speaker, crossover or terminal indicates a positive (+) connection. Crossover polarity is indicated by (+) for positive and (-) for negative or common.

It is important to observe correct polarity when making connections in all aspects of loudspeaker assembly. If wire connections are made improperly the loudspeaker will be out of phase and performance will suffer. For this project, wiring connections will always be positive (+) to positive (+) and negative common (-) to negative common (-).

Crossovers should be screwed onto the back baffle of the enclosure away from speaker magnets.

4

Two 50 Watt L-Pads.

L-Pads

An L-Pad is an adjustable electrical device used to control the volume level of a loudspeaker. In this project, a 50 watt L-Pad will be used to control the volume of the tweeter so that it can blend with the woofer to your preference. The L-Pad enables you to adjust the tweeter to your taste. The L-Pad will also allow you to adapt the loudspeakers to a wide variety of room acoustics. When you move a loudspeaker from room to room or change its position within a room, an L-Pad will enable you to adjust the way the loudspeaker sounds.

An L-Pad generally has three wiring posts numbered 1, 2 and 3. Number one is negative (-) or common. Number two is output and number three is the input connector. L-Pad wiring connections will be made between the high section of the crossover and the tweeter.

Although the connection of the L-Pad between the high section of the crossover and the tweeter is simple, it is the most complicated wiring aspect of this project. If you want to solder, we suggest that you initially pinch or crimp wires to all the connections between the L-Pad, crossover and tweeter and then test the L-Pad before soldering to make sure all the connections are correct.

Note: *A resistor does the same thing as an L-Pad. It affects the volume of a loudspeaker. There is no audible difference between an L-Pad and a resistor. However, a resistor can't be adjusted. It's fixed at one level. Trying to wire the correct value resistor for volume control into a crossover so that the speaker sounds precisely the way you like it can be a frustrating experience and a real hassle if you plan on moving the loudspeakers. Besides, L-pads are fun to play with.*

Over a period of years, L-Pads may wear out and sometimes they have a tendency to crackle a little when you adjust them—but only when you adjust them. This is nothing to be concerned about. Once an L-Pad is set to the level you prefer, it generates no crackle or noise at all. Furthermore, in the event that an L-Pad does wear out, it can be replaced. L-Pads ensure loudspeaker flexibility in any acoustic environment.

5

L-Pad

L-Pad Connections

The negative or common (ground) (-) terminal on the L-Pad is marked number 1. It is connected to the negative or common (-) tweeter terminal on the crossover and is split to also connect to the negative common terminal on the tweeter itself. One crimped female connecter is used to hold both wires to form the split connection.

The output terminal on the L-Pad is marked number 2. It is connected to the positive tweeter terminal on the crossover.

The input terminal on the L-Pad is marked number 3. It is connected to the positive terminal on the tweeter.

Note: *Check your L-Pad to make sure the polarity of the connecting posts corresponds to the numbers indicated in the diagram in section six. If your L-Pad connecting posts indicate different negative or common (-), input (+), output (+) numbers, simply make the appropriate connections. Some L-Pads have four terminals. On a four-terminal L-Pad, two of the terminals should be negative or common (ground) (-).*

6

Detail – L-Pad Wiring Diagram

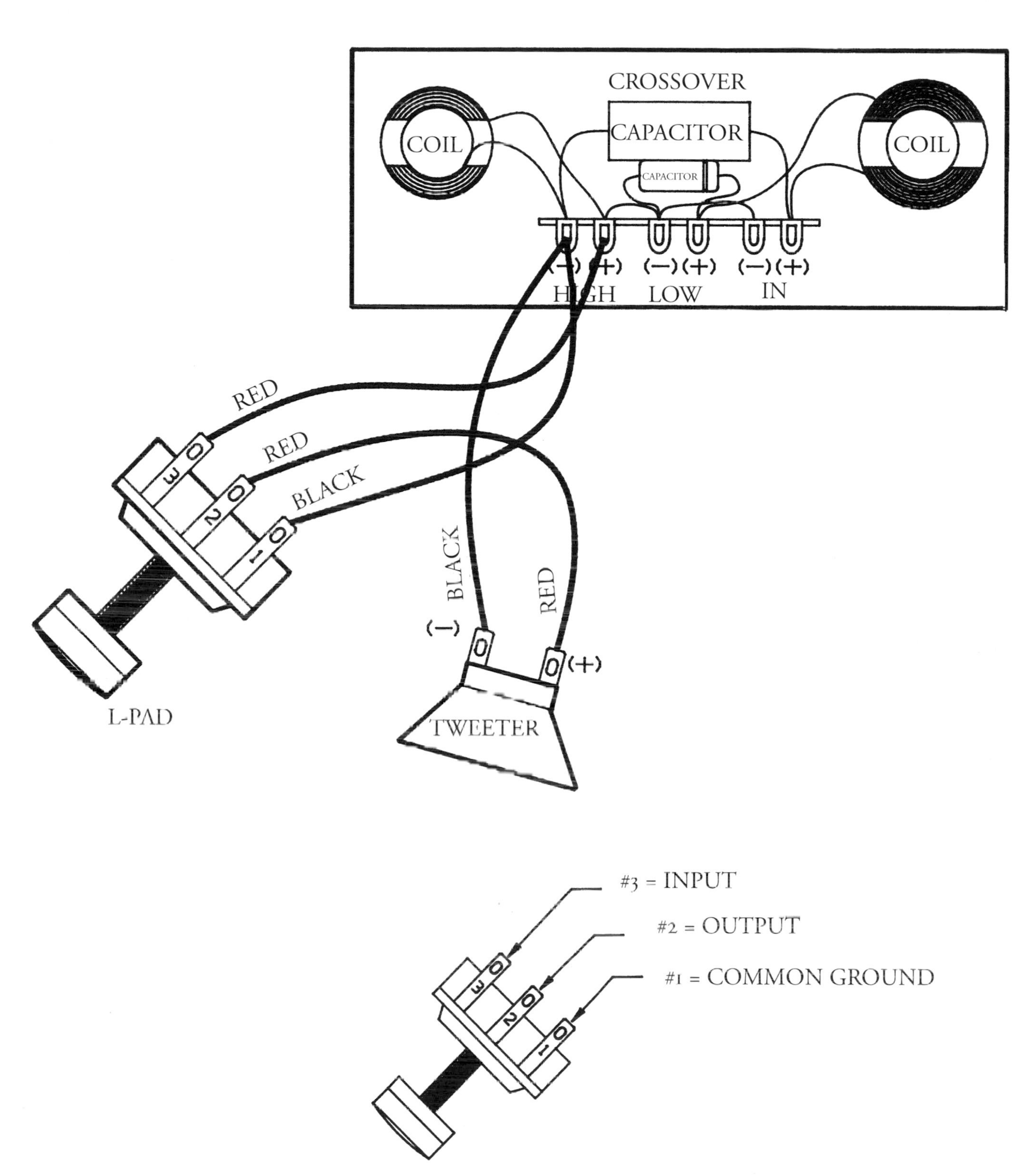
CROSSOVER
COIL
CAPACITOR
CAPACITOR
COIL
(−) (+)
HIGH
(−) (+)
LOW
(−) (+)
IN
RED
RED
BLACK
BLACK
RED
(−)
(+)
L-PAD
TWEETER
#3 = INPUT
#2 = OUTPUT
#1 = COMMON GROUND
L-PAD

2" x 5" Port Tubes.

Port Tubes

A port tube is a plastic pipe inserted into a loudspeaker enclosure. Port tubes enhance the low frequency sound characteristics (bass response) of a loudspeaker. Loudspeakers with port tubes are called bass reflex loudspeakers. For this project, a port tube will be inserted through a hole in the front baffle of each loudspeaker enclosure.

The length and diameter of a port tube is determined mostly by the volume of a loudspeaker enclosure. For our project, the port tube will be 2" in diameter and 5" long. We recommend that you use port tubes made specifically for loudspeaker construction. These tubes have a flange which makes them easy to attach to the speaker enclosure. They are available through catalogues. If you can't find the exact length port tube, cut it to size with a hacksaw.

Note: *The length of the port tube may be better if it's a little shorter or a bit longer, depending on the woofer you select. The 2" diameter will work no matter what the length. It may be worth experimenting with the length of the port tube. Although the differences may be subtle, a loudspeaker can be tuned by ear with a port tube. The 5-inch length suggested is an average worked out for the volume of the speaker enclosure described in this manual and will work well with almost all 6-1/2" woofers.*

Polyester batting and polyester fill.

Acoustic Damping Materials

Acoustic damping material is used to line inside surfaces of a loudspeaker enclosure. Acoustic damping materials used in loudspeakers are often synthetic fibers, similar to the stuffing in a pillow.

Damping a loudspeaker enclosure dramatically affects the way a speaker sounds. Both the amount and placement of damping materials within an enclosure are important. A loudspeaker enclosure without any damping material tends to have a "peaky" or hollow sound (boxiness). A loudspeaker with too much damping material will sound muted. Experiment with the amount and placement of damping material.

We suggest that you use polyester batting alone or in combination with polyester fiberfill for damping your enclosures. Polyester is easy to handle and can be purchased at fabric stores. Staple damping material to the inside of the enclosure with a heavy duty staple gun. Make sure the staples are secure or they will wind up attached to your speaker magnets.

Note: *Damping material should not be placed on the inside surface of the front enclosure baffle that holds the loudspeakers. Make sure damping material does not block port tubes. Experiment with damping material in order to get the loudspeaker to sound the way you want. Speaker enclosure damping is subjective. What sounds best to you may not appeal to someone else. Sometimes just lining the top third of the enclosure with polyester batting works best.*

The following photos illustrate effective examples of enclosure damping. Remember, the loudspeaker will sound different when you change the amount and placement of damping material within the enclosure.

Enclosure fully damped with polyester batting and fiberfill.

Enclosure lined with polyester batting.

Top-third enclosure lined with polyester batting.

Back-panel of enclosure loosely damped with polyester batting over terminal cup, cross-over and L-Pad.

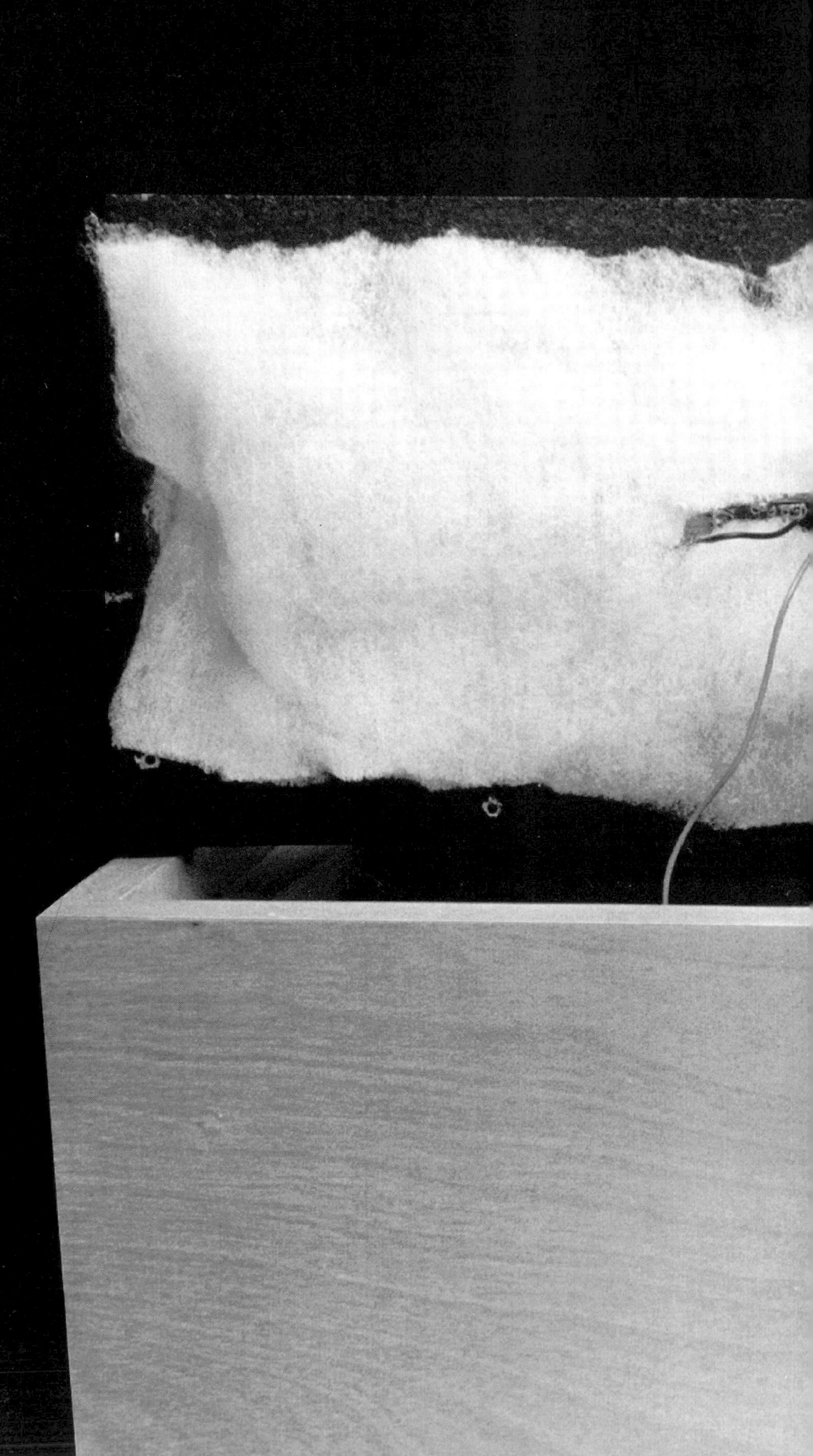

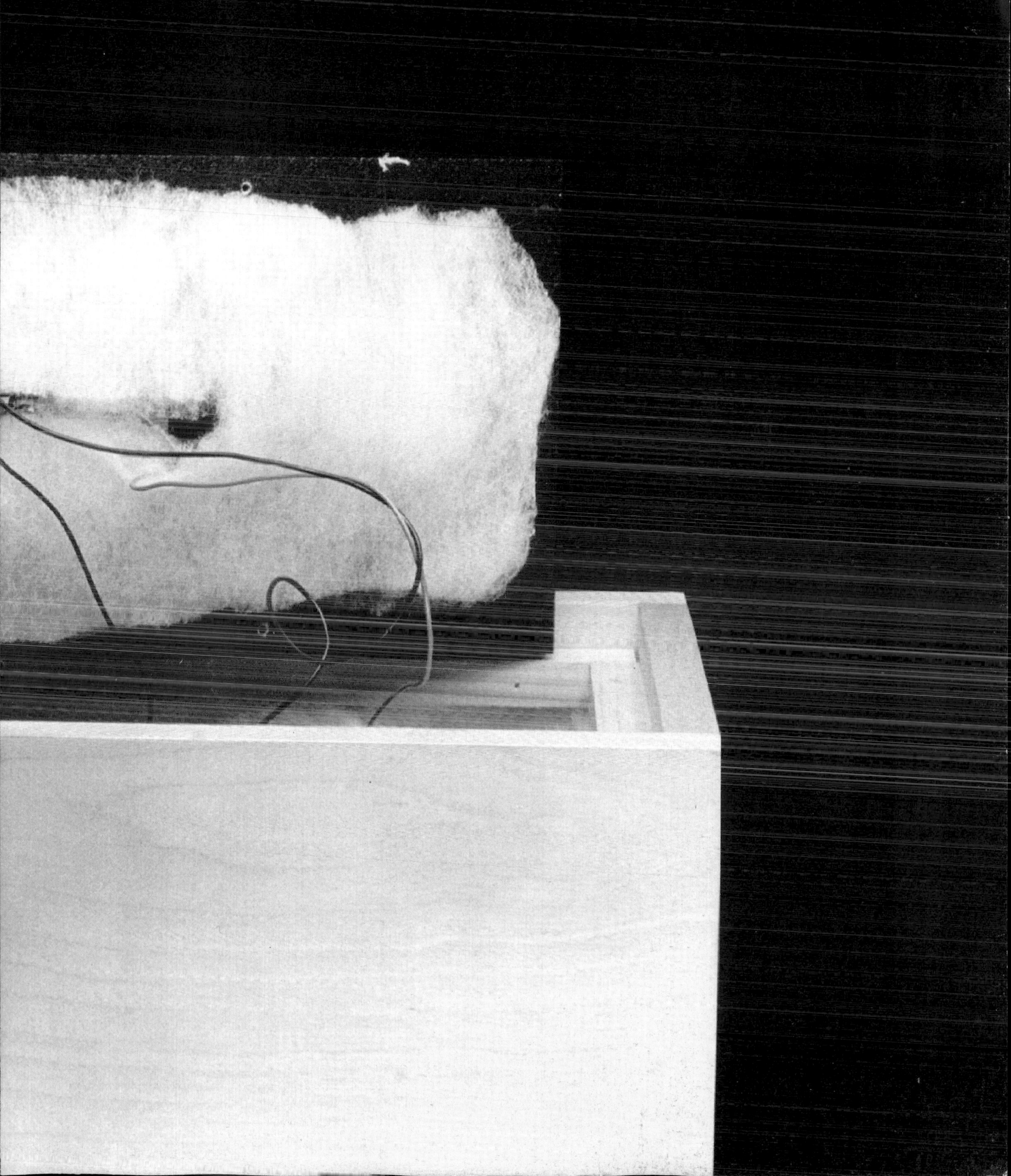

Audax 1" Soft Dome Tweeters (Model #TWO25A1).

Speaker Suggestions

We suggest you use dome tweeters. Dome tweeters have superior high frequency dispersion characteristics.

Audax of America offers a one-inch 8 ohm soft dome tweeter that is superb (Model #TWO25A1). It has a low-resonance frequency and high-power capability. This tweeter will work well with crossovers discussed in the manual.

Audax also offers a number of 6-1/2" woofers with die-cast chassis. The Aerogel, Carbon Fiber, Kevlar and Treated Paper Cone woofers are among some of the fine woofers this company offers. The die-cast chassis paper cone was used for illustration.

Companies that sell loudspeakers and components are listed in the back of this manual.

Audax 6-1/2" Paper Cone Woofer (Model #HM170GO).

10

Recessed terminal cups. Press-fit type and binding-post type with knurled knobs.

Terminal Cups

A terminal cup is an electrical connector attached to the back panel of the speaker enclosure. It is used to connect the loudspeaker to the amplifier or receiver and has positive and negative terminals. The inside of the terminal cup is connected to the positive (+) and negative (-) "IN" connectors on the crossover. In this project, we will be using what are called recessed terminal cups. They are available at electronics stores or through catalogues.

There are two common types of recessed terminal cups–the kind with press-fit push-buttons and the kind with binding posts and knurled knobs. Either type works well.

Terminal cups have a red connector and a black connector. Red is for the positive (+) connection and black is for the negative or common (-) connection. Make sure to observe the correct polarity when connecting wires from your terminal cup to your amplifier and when wiring internally to the crossover.

Back side of a recessed terminal cup.

II

Solderless Female Connectors (3/16" and 1/4").

Wiring with Solderless Female Connectors (Quick Disconnects)

For the beginner new to loudspeaker construction, there are a number of reasons why using solderless female connectors is advantageous for wiring your loudspeakers. First of all, if you make a mistake, it's no big deal to correct the error; all you have to do is unplug the improper connection. However, if you start out using solder and make a wrong connection, it's not as easy to correct things. Solderless female connectors (female quick disconnects) are available at electronics and many hardware stores.

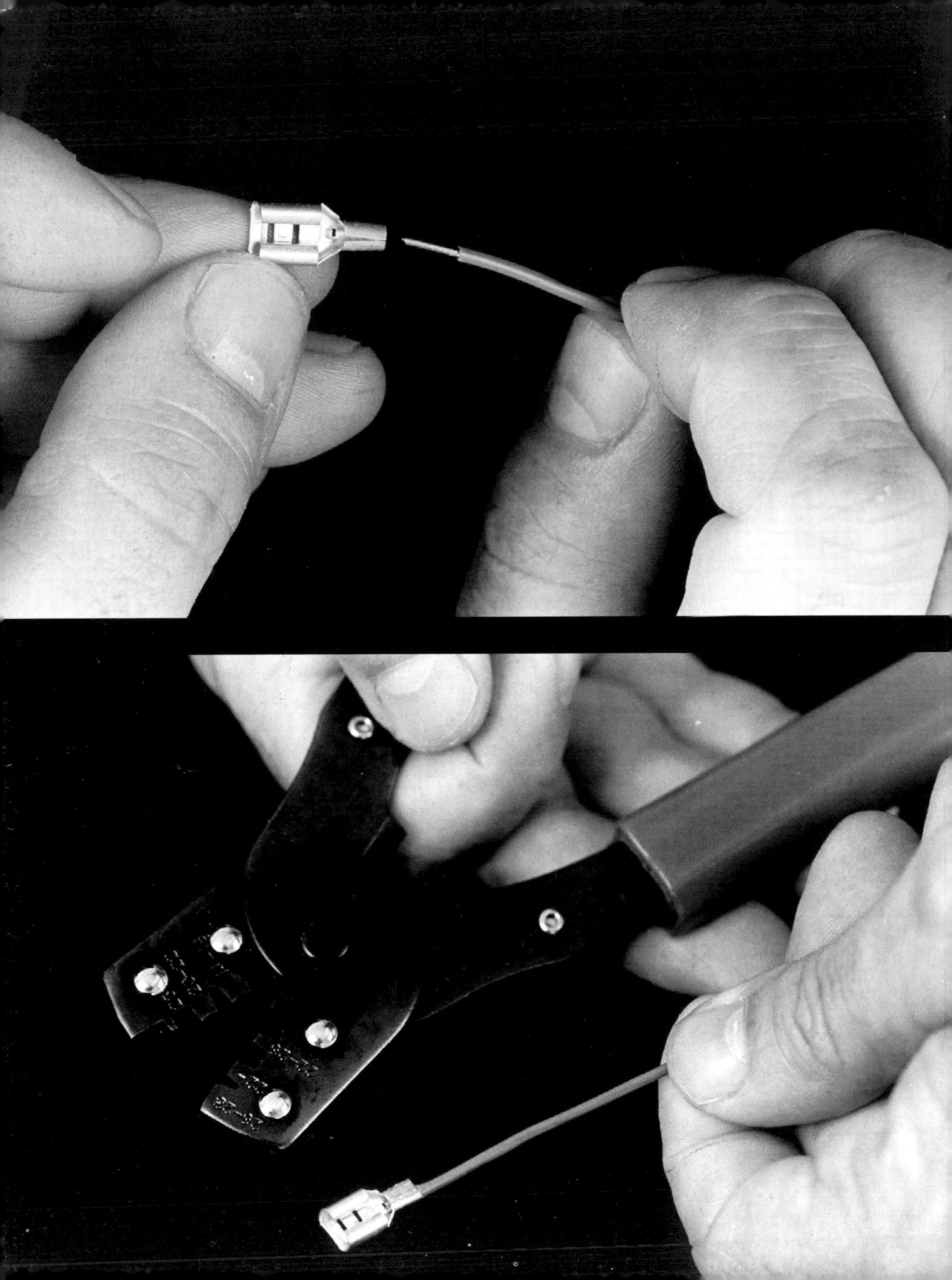

(Upper left) Stripped wire inserted into neck of female connector.
(Lower left and above) A crimping tool is used to pinch a connector closed.

3/16" female connector for a woofer hook-up (upper left) and 1/4" connector used for a terminal cup (upper right). Connecting posts on components may vary in size. Check sizes before selecting female connectors.

12

Parts and Materials

1. Two 6-1/2" woofers, 8 ohm

2. Two tweeters, 8 ohm

3. Two two-way crossovers, 8 ohm

4. Two 50-watt L-Pads, 8 ohm

5. Two 2" diameter x 5" long plastic port tubes

6. Two recessed terminal cups

7. Twelve feet (12') of braided or solid copper connecting wire with red insulation for positive (+) electrical connections.

8. Twelve feet (12') of braided or solid copper connecting wire with black insulation for negative or common (-) (ground) connections.

9. One sheet of 3/4" MDF, particle board or plywood. You won't need the whole sheet but it may be difficult to squeeze what you need out of a half sheet. It may be handy to have some extra material. Some lumber yards will cut the material for you. The type of construction shown in this manual is butt joint. If you wish to build a pair of butt joint enclosures you will need 3/4" thick material cut to these sizes:

   Top & bottom pieces – 4 required – 7-1/2" x 13-3/4"
   Side panels – 4 required – 19-1/2" x 13-3/4"
   Back panels – 2 required – 18" x 7 1/2"

7-1/2" x 18" front baffle is 1" thick.

10. If you wish to use the butt joint construction illustrated in this manual, you will also need 2 front baffle pieces cut from 1" thick MDF, particle board or plywood. Cut to 7-1/2" x 18".

11. Eighteen feet of 3/4" square baluster for framing butt braces around inside of enclosure. Available at lumber yards. Any equivalent will do. (See construction plans for cut lengths.)

7-1/2 x 18" back panel is 3/4" thick.

12. Two feet of 1-1/4" or 1-3/16" square baluster for the internal enclosure brace. Available at lumber yards. Any equivalent will do. Brace should be cut to the internal width of your enclosure. The enclosure illustrated in this manual has an internal width of 7-1/2". However, the width of your enclosure may vary somewhat. Measure before you cut the brace. The baluster brace is set 6-1/2" down from the top of the enclosure and 6-1/2" back from the inside of the front baffle. Cut the brace so it fits tightly. Sand or cut opposite corners of the brace so that it can be wedged into the enclosure. After you fit the brace, the ends of

3/4" baluster for framing interior butt braces of enclosure.

the brace should be glued to the side panels. The brace should fit tightly enough to be self-clamping. The purpose of the internal brace is to minimize box vibrations or box resonance. (The brace helps keep the box from making noise.)

13. Two pieces of 3/8" plywood cut 9" x 19-1/2" for loudspeaker grills. The centers of these blanks will be cut out with a jigsaw to form the frame of a loudspeaker grill. Some people like the way their speakers sound and look without speaker grills so this element of your loudspeakers can be considered optional.

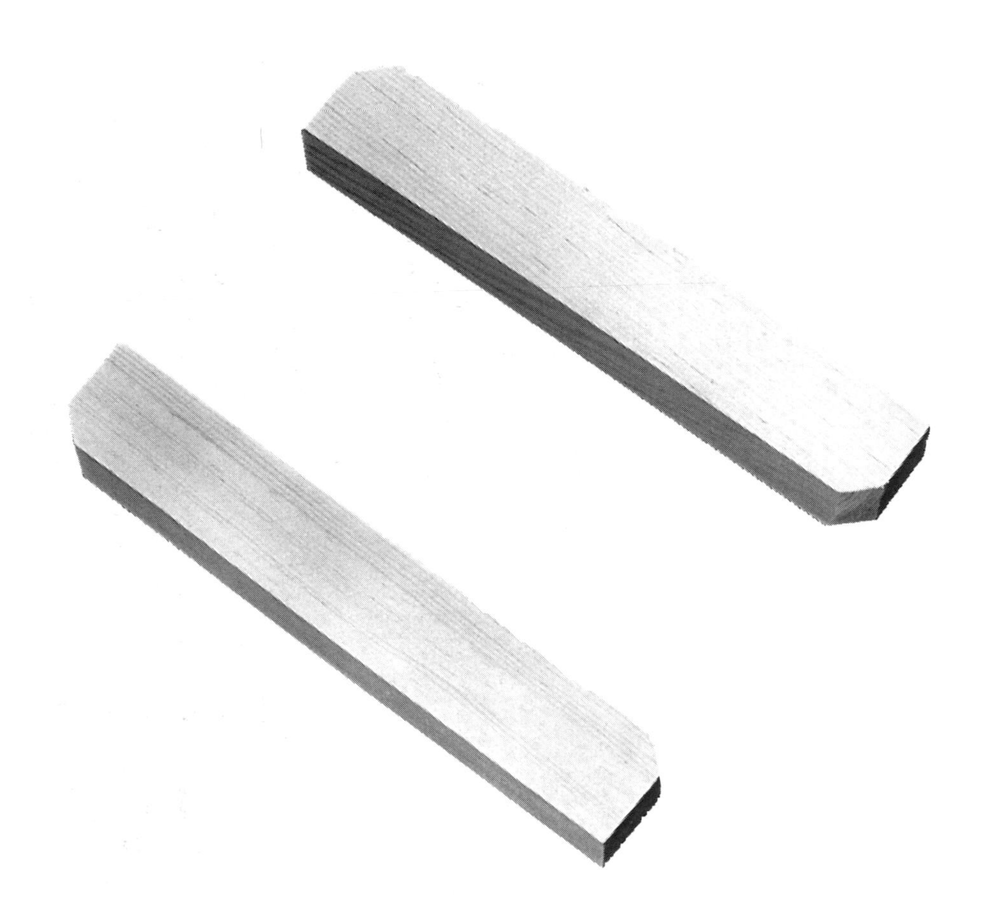

1-3/4" baluster for internal brace. Opposite ends are trimmed to allow wedge fitting.

14. One square yard of fiberglass window screen or loosely woven synthetic fabric of your choice for stretching and stapling around the back of the loudspeaker grill. Avoid natural fiber fabrics like linen, cotton and wool because they tend to shrink and can warp the grill.

15. Two feet of adhesive-backed loop and hook strips for securing the speaker grill to the enclosure. In addition to the adhesive on the loop and hook strip, we suggest that you also use staples to reinforce adhesion.

Enclosure with brace installed.

16. High-quality carpenter's wood glue.

Important: *We suggest that you do not glue the front baffle or back panel of the enclosures until you are sure everything is just the way you want it and you have familiarized yourself with your components. Once the front baffle and rear panel are glued, any changes will have to be made through the woofer opening. Some people just screw the baffle in place so they can replace the baffle if they want to make a change in speaker configuration.*

Note: *Rattling in a well-made loudspeaker enclosure is rare. Rattling in a removable baffle can usually be cured by either tightening the screws, shimming the baffle joint with felt or gluing the baffle to the enclosure. If the baffle fits snugly, it won't rattle.*

17. A small can of semi-gloss black paint (or other color of your choice) for painting the front baffle. Some people prefer a veneered baffle with the same finish as the rest of the loudspeaker enclosure.

18. A can of polyurethane or other preparation you may prefer for finishing the outside of the enclosures.

19. Two dozen small 1/2" Phillips head screws for fastening crossovers inside the speaker enclosures and terminal cups to the backs of the enclosures. (You probably won't need them all, but it's a good idea to have a few extra). Bring components to a hardware store for sizing screws. Pre-drill for all screws.

20. Two dozen roundhead 3/4" screws. These are for fastening the speakers to the front baffle. Make sure you have the right diameter screws for your speakers.

21. A box of drywall screws or multi-use screws for fastening all glued joints of the enclosure together. For the materials specified in this manual, we suggest that you use #6 multi-purpose 1-1/4" screws for fastening the enclosure together and #6 multi-purpose 1-5/8" screws for fastening the front baffle. Be certain that the screws you use are not too long.

Important: *Be sure to pre-drill holes for all screws. Make sure the entry holes for the screws are almost as big as the threads. Drill through your material the length of the screw. All butt joints should be well-glued on all touching surfaces.*

MDF (medium-density fiberboard) is tough stuff. If the pre-drilled holes are too small, the heads may snap off the screws as they are screwed in. Practice pre-drilling holes for screws by screwing scraps together.

22. A tube of clear silicone rubber adhesive. Clear silicone adhesive is used to help secure the L-Pad to the rear panel of the enclosure. It can also be used to affix the port tube to the front baffle if you cut the hole a little bit too big. (See photo on the next page.) You can use it to secure the crossover to the inside of the box along with screws. Silicone adhesive can be used to line interior joints of the enclosure. However, if you build a tight box, you won't need it.

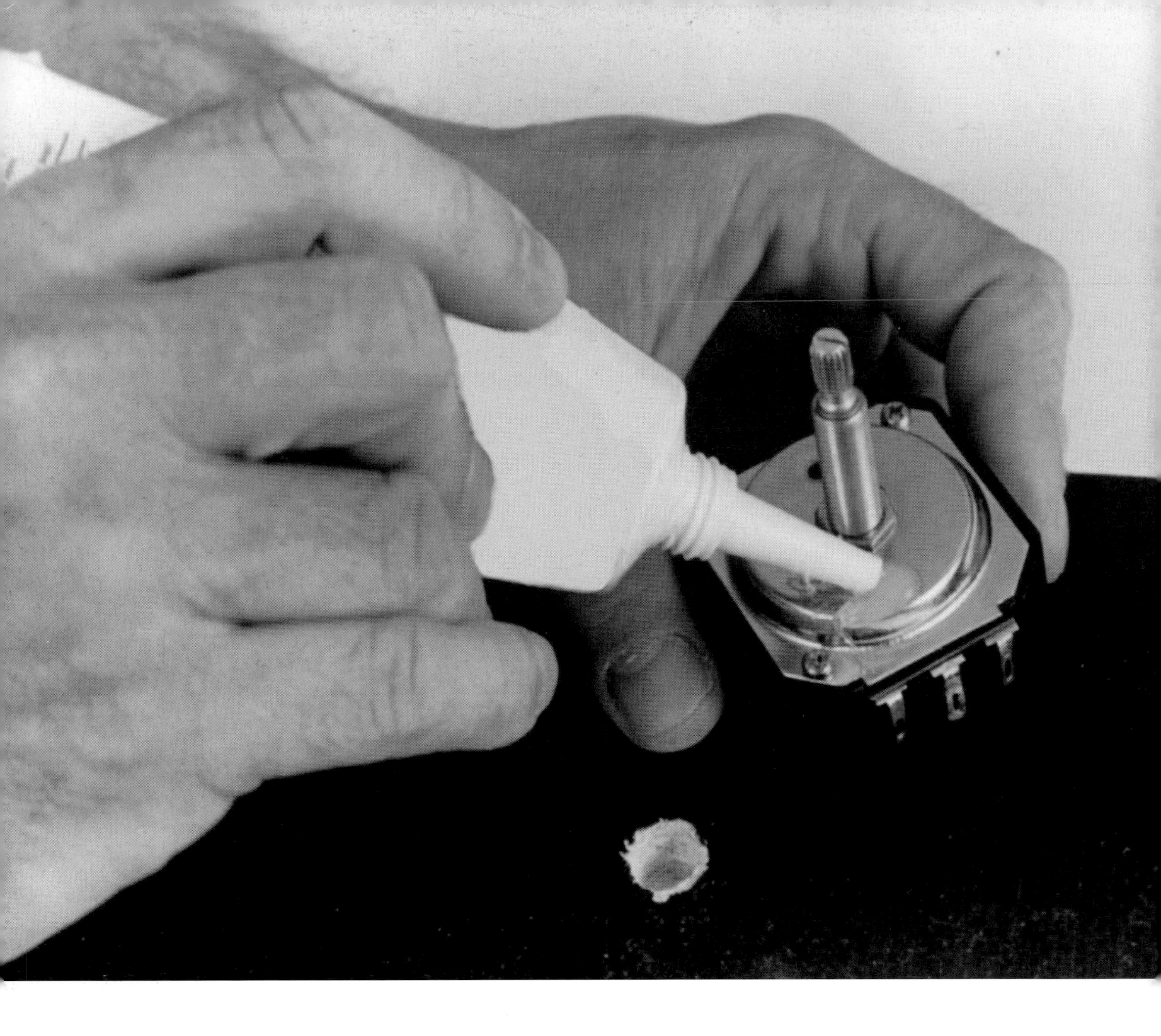

Silicone adhesive applied to L-Pad

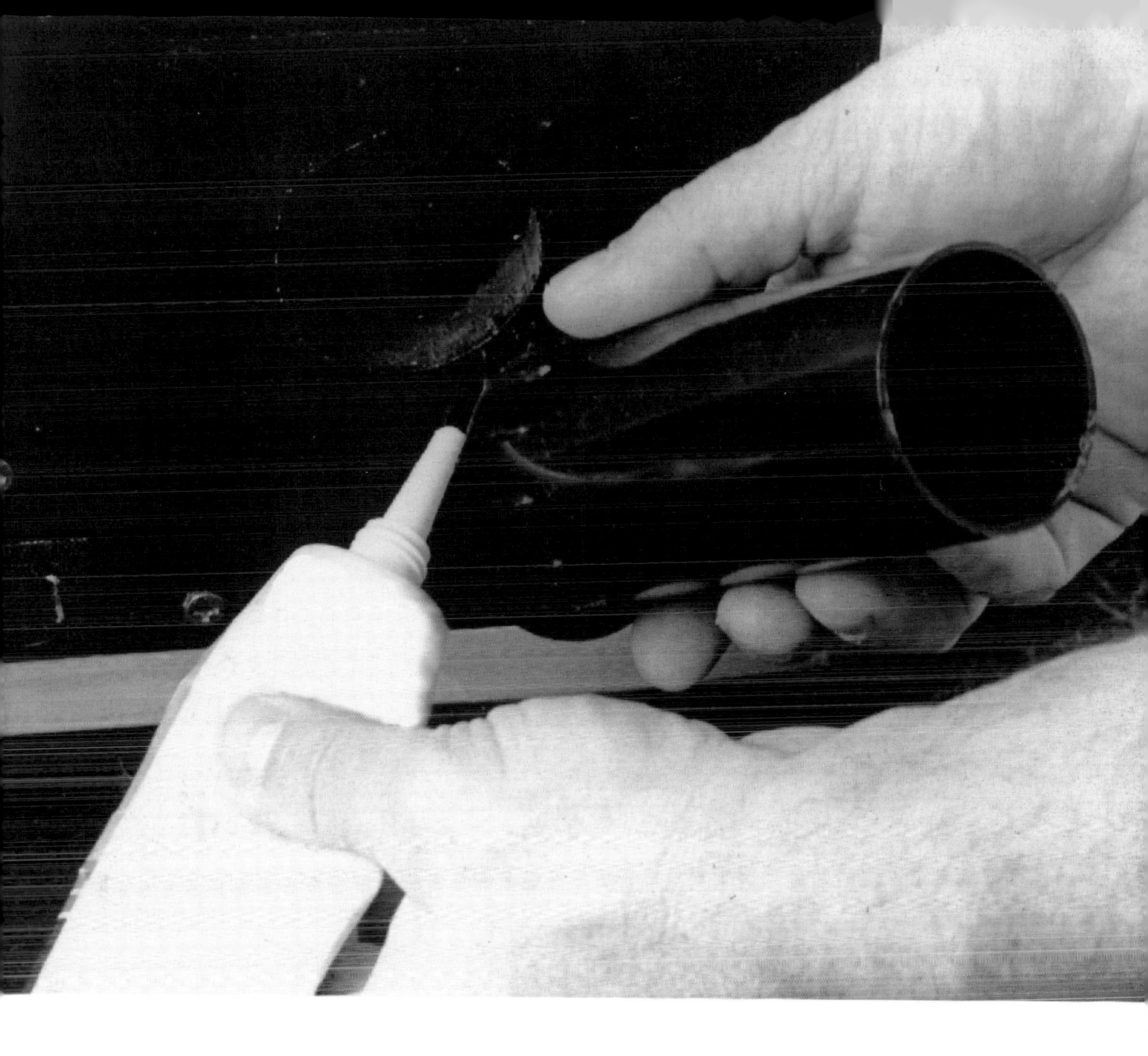

and port tube flange.

Multi-purpose screws used for enclosure construction

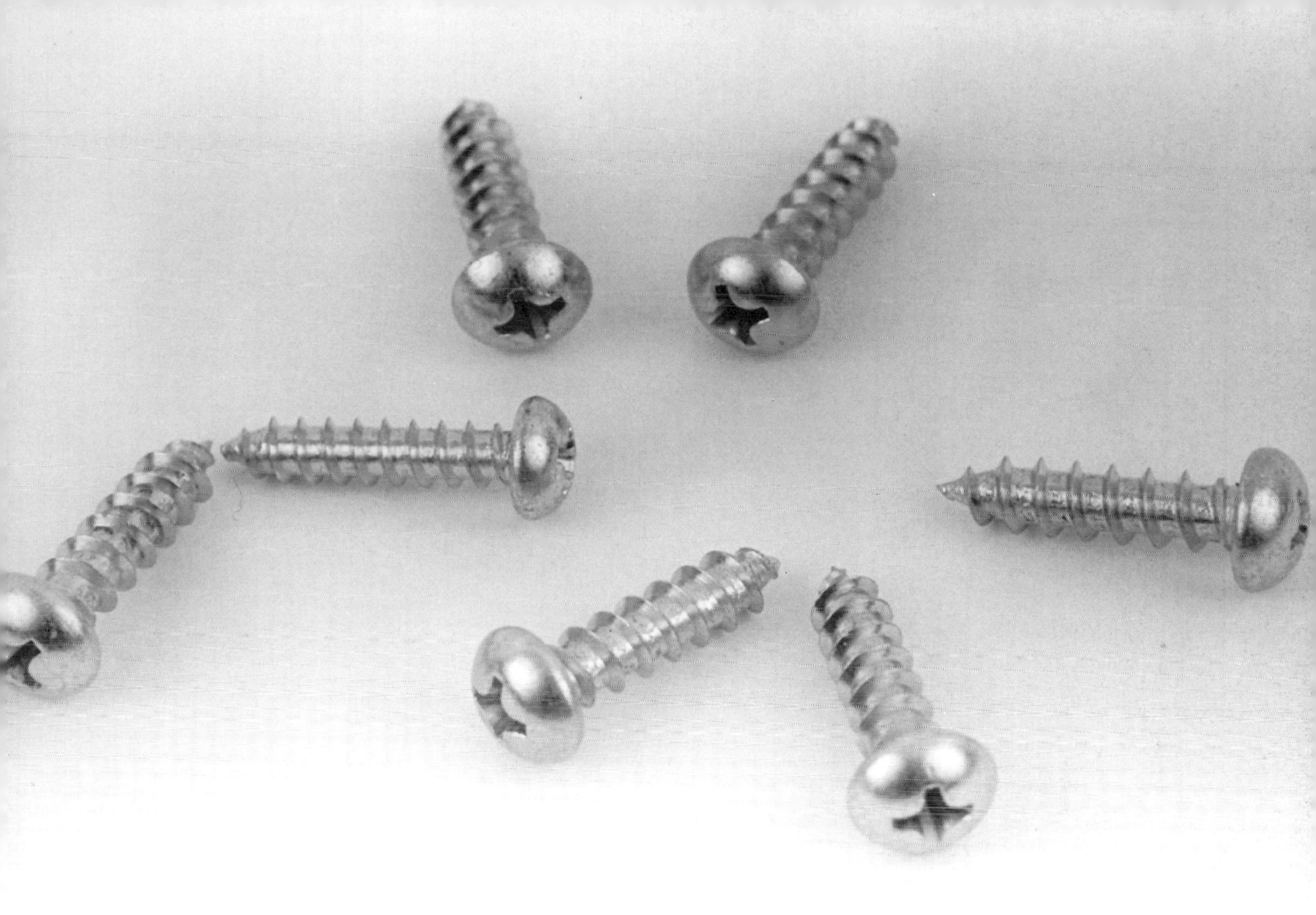

and round head screws for attaching speakers to the front baffles.

13

A hand-held power jigsaw cutting out woofer opening.

Tools

Caution: *If you are not experienced in the use of power tools, have someone who is experienced cut enclosure parts to your specifications.*

Try to keep the internal (inside) dimensions as close to specifications as possible no matter what type of construction you use.

ALWAYS WEAR EYE PROTECTION WHEN USING ALL TOOLS!

A hand-held power jigsaw.

1. *Table Saw:* A table saw will be used to cut out the front, back, sides, tops and bottoms of your enclosures.

2. *Hand-Held Power Jigsaw:* A jigsaw will be used to cut out the loudspeaker openings and port tube openings in the front baffles of the loudspeaker enclosures. A jigsaw will also be used for cutting out the recessed terminal cup opening on the back of the speaker enclosure and grill frames.

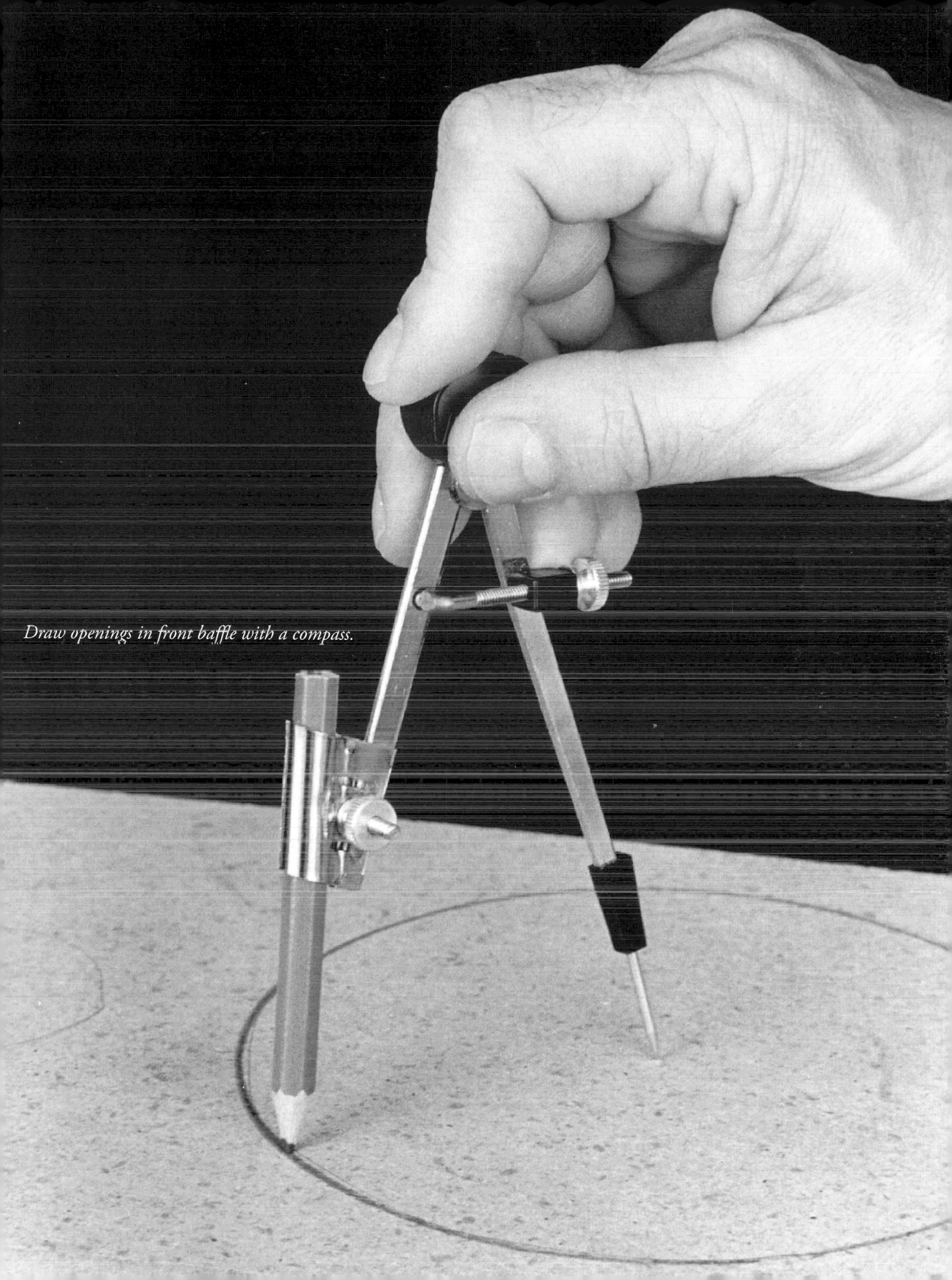

Draw openings in front baffle with a compass.

Note: *The following drawings show you how to measure a woofer and tweeter for cut-out dimensions. These drawings are not to be used for actual measurements.*

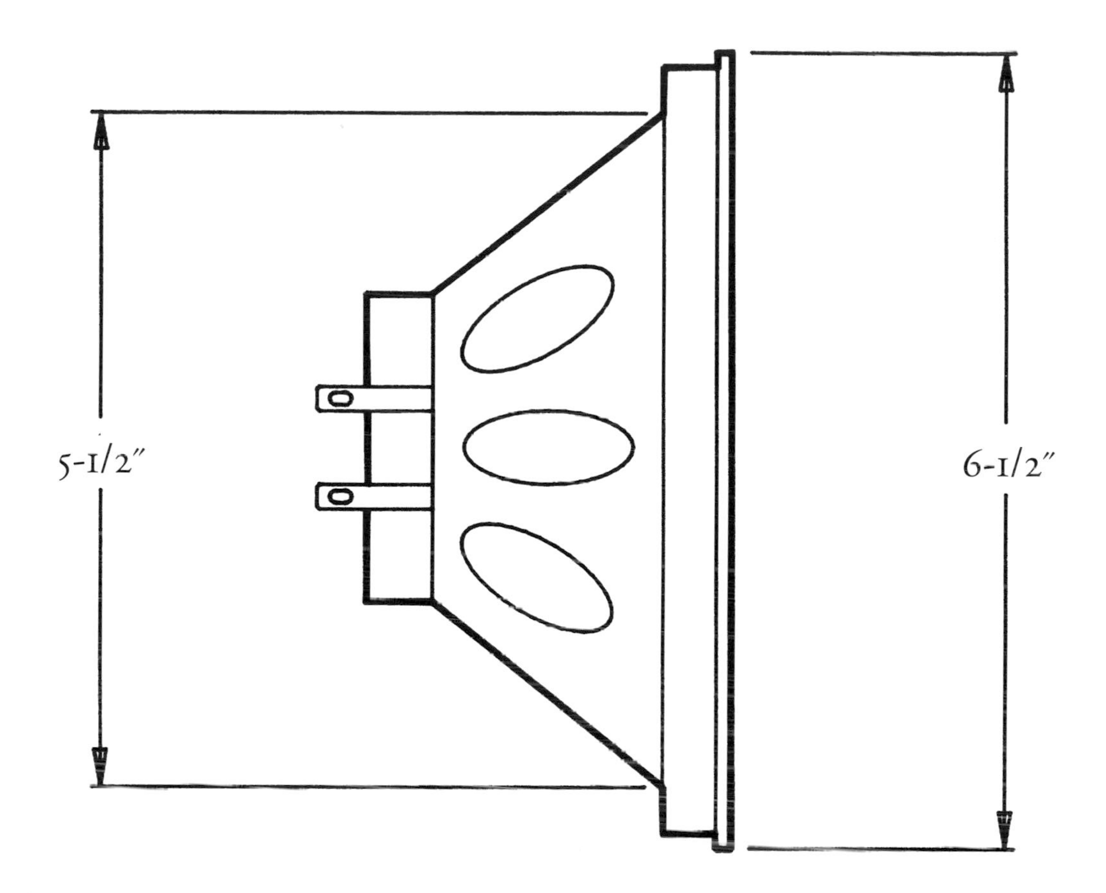

Hypothetical Woofer (side view). Cut out would be 5-1/2" in diameter.

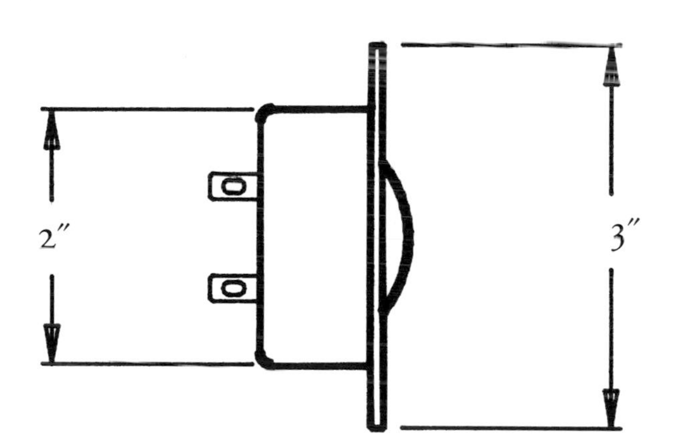

Hypothetical Tweeter (side view). Cut-out would be 2" in diameter.

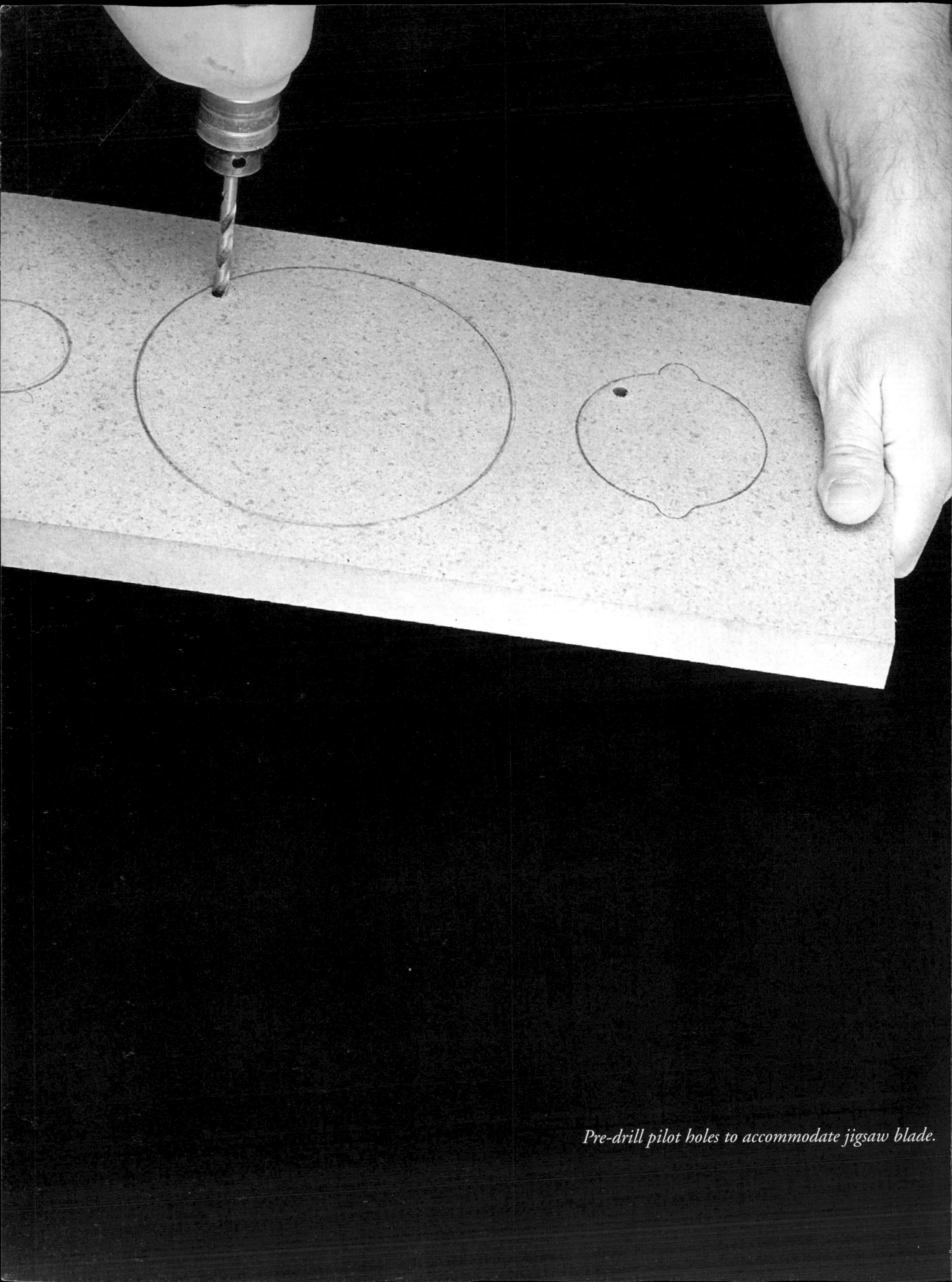

Pre-drill pilot holes to accommodate jigsaw blade.

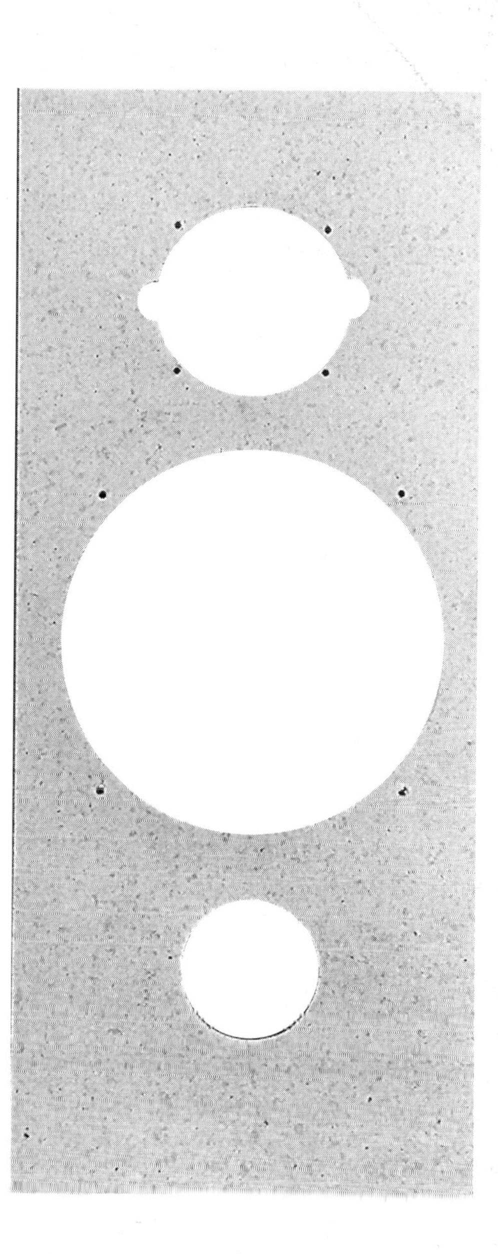

Front baffle openings cut out with jig saw.

3. *Router (optional):* A router will be used for flush-mounting your woofers to the front baffle of the speaker enclosure. Flush-mounting helps eliminate edge diffractions. If you wish to flush-mount the woofer and/or tweeter to the loudspeaker baffle, use a router for cutting a recessed edge around the cut-out opening on the baffle. Use a 3/4" or 1/2" straight bit in the router.

Caution: *If you have never used a router before, have someone familiar with using a router perform this procedure for you. Always wear a mask or respirator and eye protection when routing! It is beyond the scope of this manual to instruct anyone in the proper use of a router.*

Note: *Flush-mounting is an optional procedure. Your speakers will sound fine if you just screw them into the openings cut in the baffles with the jigsaw.*

The frames or rings around tweeters are thinner than those around woofers and they don't have to be flush-mounted. The tweeters illustrated in this manual were not flush-mounted.

If you flush-mount, you may have difficulty later on if you want to change your speakers. New speakers with different size rings may not fit.

To flush-mount a woofer in a loudspeaker baffle, place the woofer in the woofer opening and trace an outline of the outside edge of the woofer frame (ring) around the opening already cut in the front baffle. Measure the depth of the woofer frame (ring) to determine the depth of the router cut in the baffle.

Outline of woofer ring being traced in front baffle for router guidelines.

Guideline of woofer ring traced for router cut.

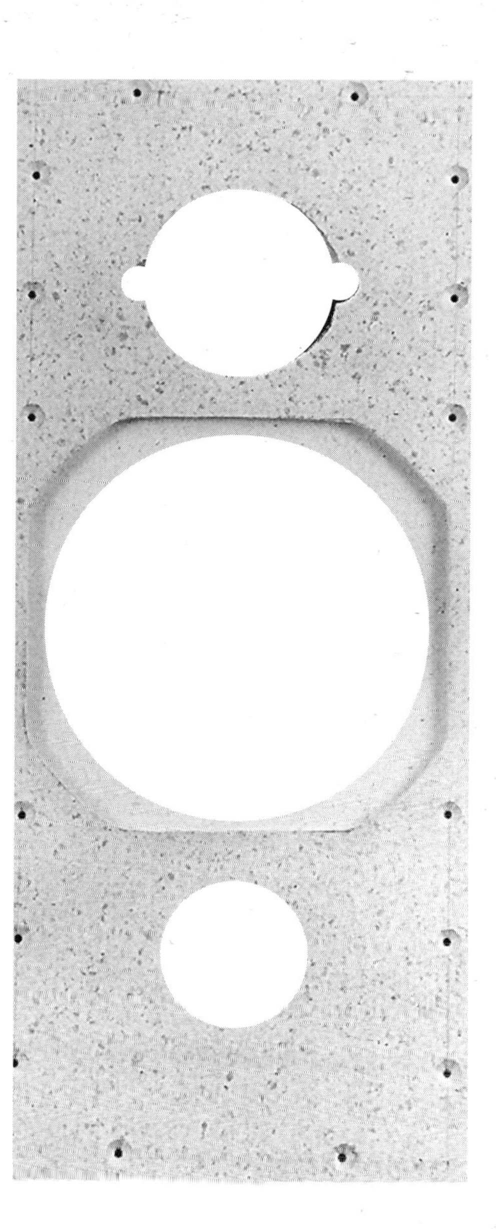

Front baffle cut to guidelines around woofer opening with router for flush-mounting the woofer.

4. *Variable Speed Reversible Electric Drill*: Make sure you have an assortment of drill bits, a countersink for setting screws flush, a Phillips head screwdriver bit and both a small and large cylindrical rotary rasp for your drill.

Large and small rotary rasps. Use the small rasp for grinding notches for tweeter terminals in the front baffle.

A countersink.

Important: *All screws should be screwed into pre-drilled holes. Measure and mark screw placement according to enclosed plans. After pre-drilling, use the countersink so exterior screws on the front baffle and rear panel will be set flush to the surface of your construction material. A small rotary rasp is great for grinding out the notches for the tweeter terminals in the front baffle. The large rotary rasp may come in handy if you need to modify the diameter of any circular opening in the front baffle. After these procedures have been accomplished, paint the front baffle and back panel black (or any color/finish you prefer). Do not attempt to paint the front baffle or rear panel when they are in the enclosure. Make sure you measure and mark placement for all screws according to measurements on the plan. If your placement of screws varies from the enclosed plans, make sure they do not intersect at any point.*

Using the countersink for setting screws flush [illegible]te: flush-mount cut-out for woofer).

Use a Phillips head screw bit for screws. (Make sure all holes have been pre-drilled.)
Drive screws carefully. Going too fast with a screw bit can damage the baffle or break the screw.

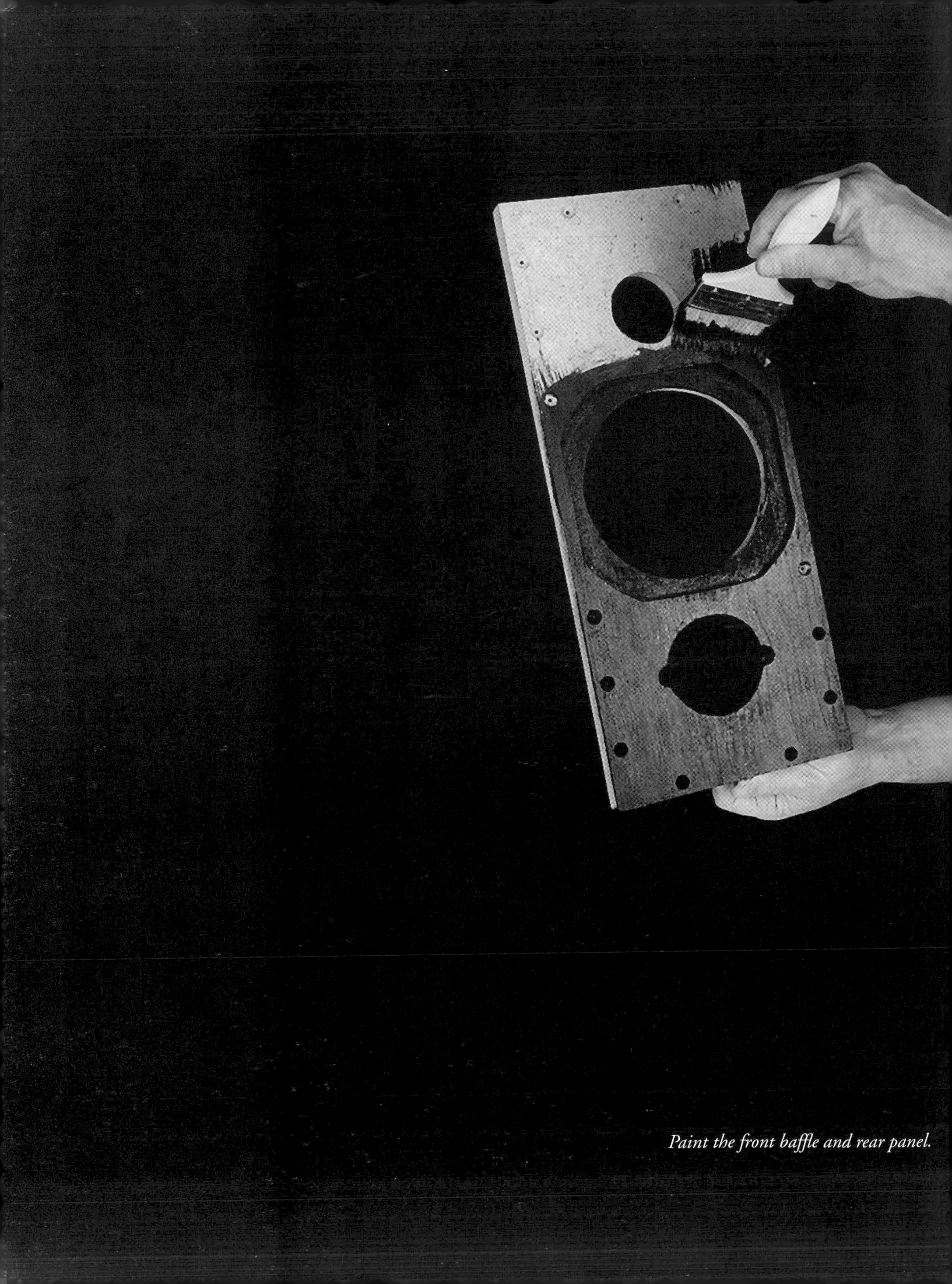

Paint the front baffle and rear panel.

Completed front baffle ready for final installation.

Wiring tools from left to right.

5. *Diagonal Cutting Plier:* Used for cutting wire.

6. *Needle Nose Plier:* Used for bending, fitting and pinching wire.

7. *Wire Stripper:* Used for stripping insulation off wire.

8. *Crimping Tool:* Used for crimping electrical connectors.

Note: *Practice crimping female connectors to wires.*

9. *Hand-Held Phillips Head Screwdriver:* Make sure you have a screwdriver small enough to allow you to work inside the enclosure.

10. *Paintbrushes:* For applying paint and/or other finish to your enclosures.

11. *A Soldering Iron and Solder (Optional):* If you wish to solder we suggest you use solder only on the crossover. You may wish to change the speakers and/or crossover at a later time. It's a lot easier to remove crimped female connectors than solder, especially from loudspeakers.

Note: *Use rosin core solder only. Make sure the tip of your soldering gun is clean. Clean the tip of the soldering gun with sandpaper before you plug it in. Before you solder, make sure all electrical connectors are clean. Pinch all wires to connections first by bending the wire around the connector tightly with a needlenose plier.*

Important: *Test all connections by listening to the loudspeakers with wires pinched onto connectors before you solder. Solder from the top down*

Caution: *Always wear eye protection! Soldering guns and melting solder are very hot. Never leave a room with a soldering iron plugged in. Always keep hot soldering iron tips and parts away from insulated wires. Never use a soldering iron near anything flammable! Always keep a soldering iron in its cradle on top of a non-flammable tray or receptacle! Always use adequate ventilation! Keep soldering guns and solder away from children and pets!*

12. *Compass:* Used for drawing circles for speaker openings.

13. *A Ruler and/or Tapemeasure.*

14. *Scissors.*

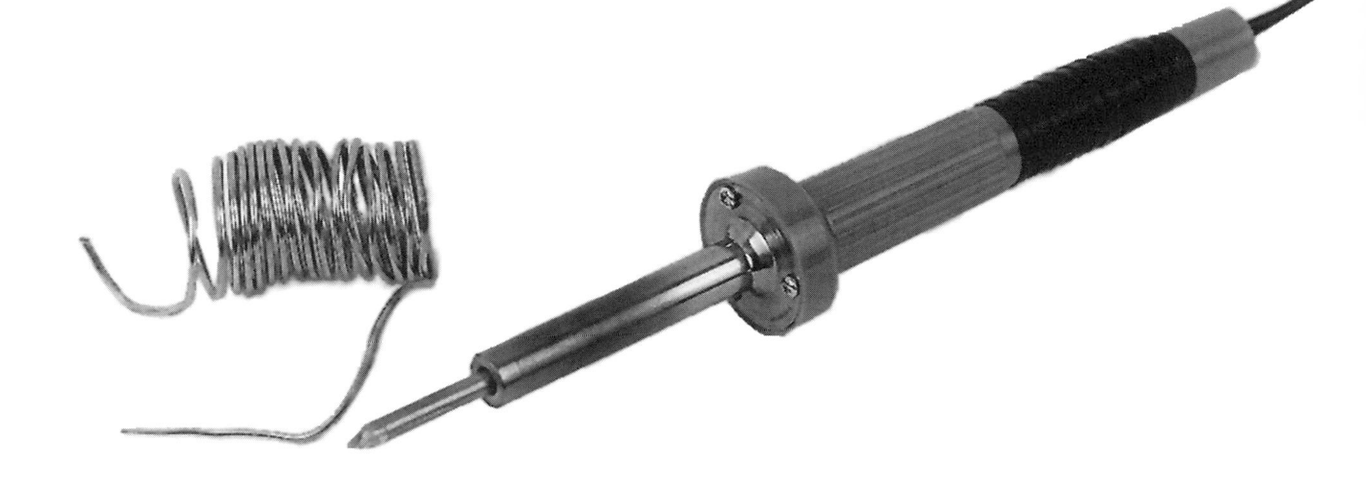

Rosin core solder and soldering iron.

15. *Staple Gun:* Make sure the staple gun is powerful enough to drive staples into the material you will be using. Staples will be used to hold damping materials in place and for the screen or fabric that will be stretched onto the back side of the speaker grill. Staples will also reinforce adhesion of loop and hook fasteners to the back of the grill and front of the speaker enclosure.

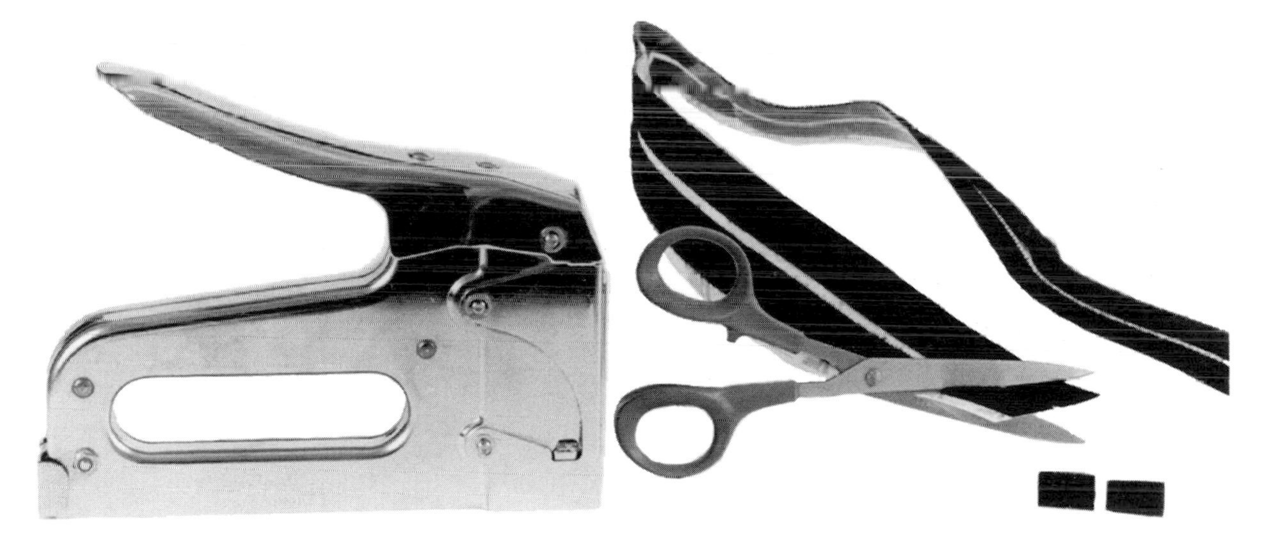

A staple gun, scissor and loop and hook strip.

14

Enclosure Plans

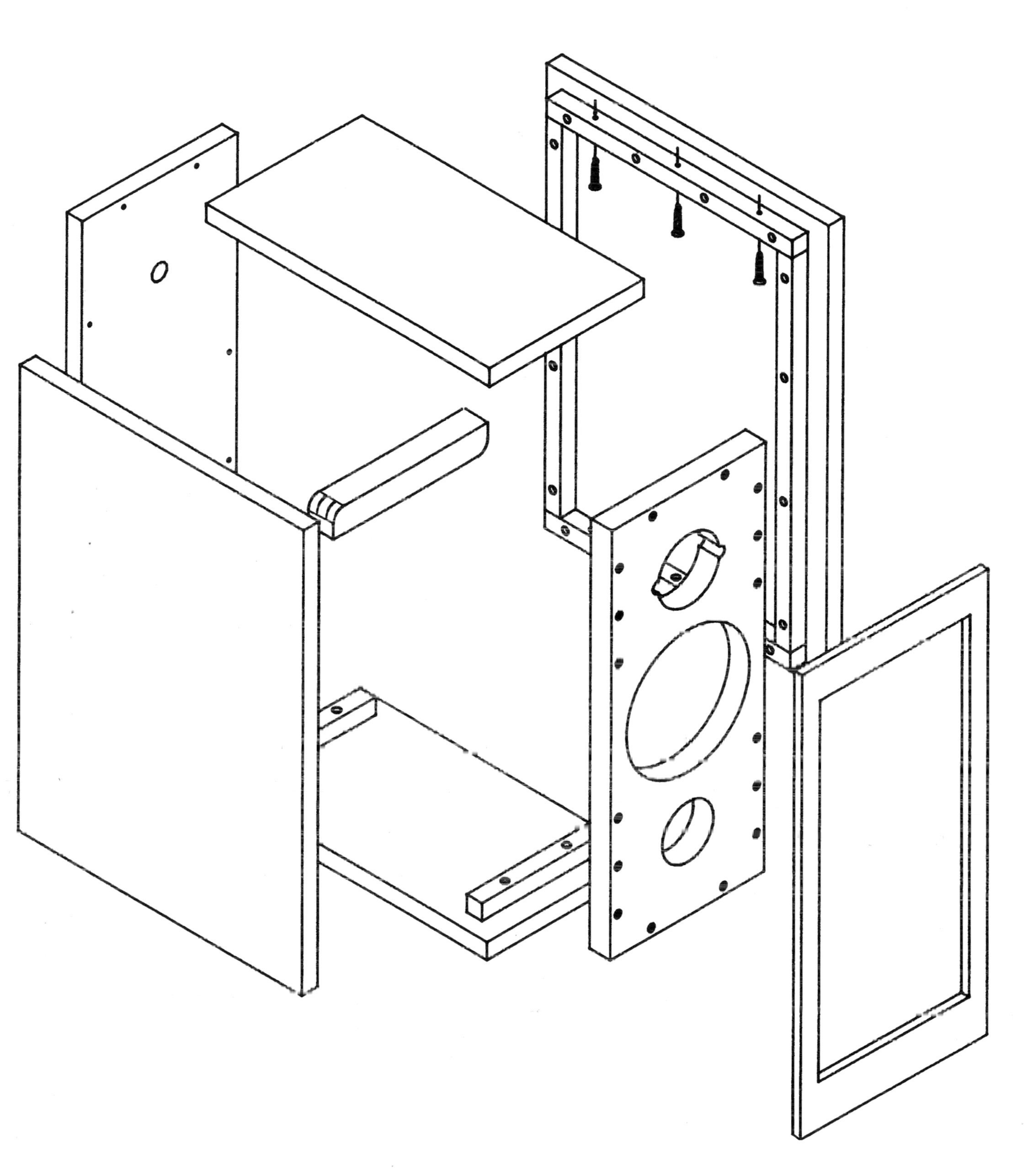

Exploded View

Material 1" thick MDF (medium-density fiberboard), particle board or plywood.

Size 1" x 7-1/2" x 18"

Note: *1. Lay your components out before cutting openings. Diameters of openings will vary according to what components you use.*
2. Draw circles with a compass. Trace around port tube with a pencil.
3. Measure carefully and check before you cut.
4. All drilled holes are 3/8" in from edge of baffle.

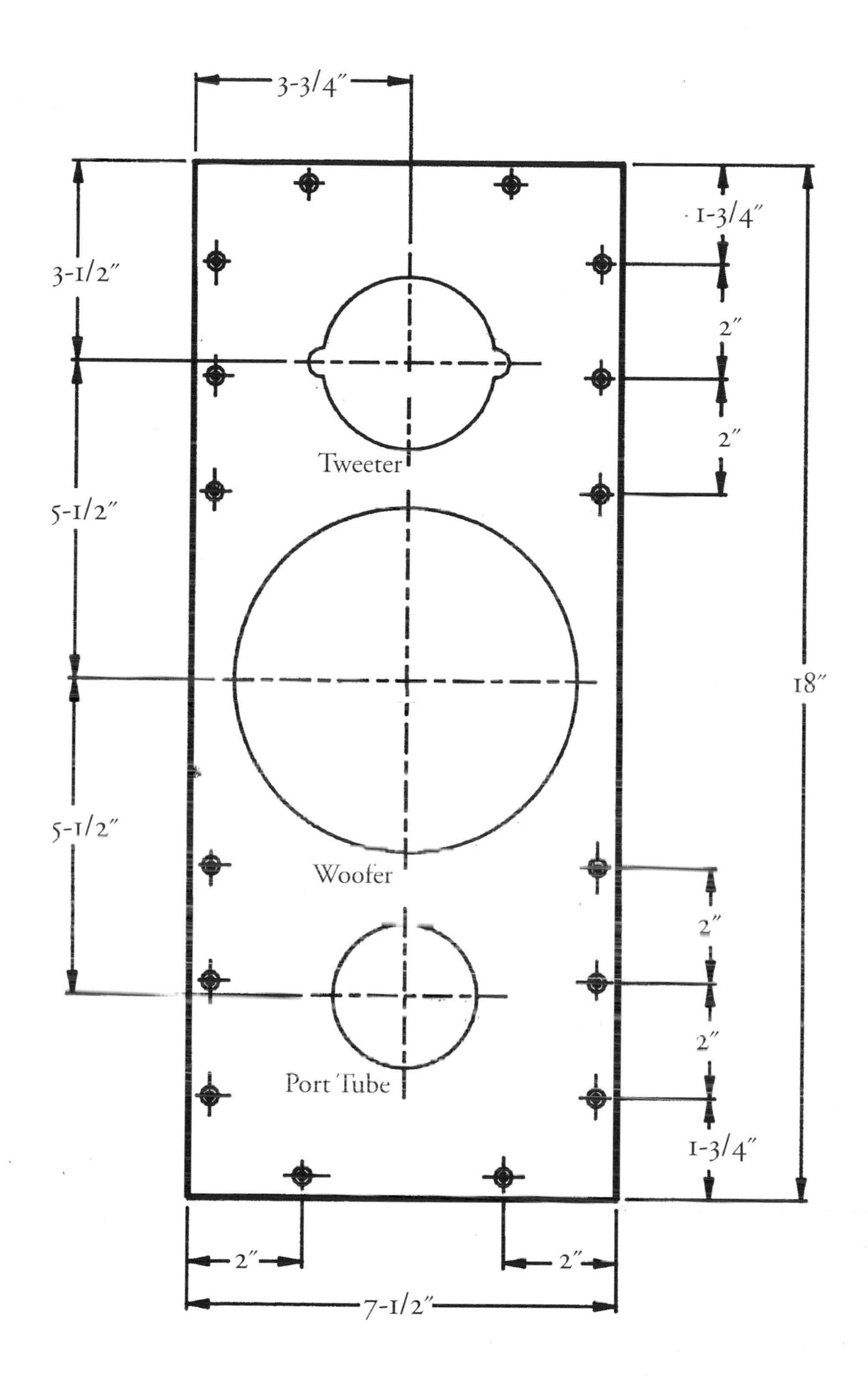

Front Baffle

Material 3/4" MDF (medium-density fiberboard), particle board or plywood.

Size 3/4" x 7-1/2" x 18"

Note: *Remove knob and/or plastic plate from the L-Pad. Measure the diameter of the L-Pad shaft and drill or cut out correct diameter hole. Recessed terminal cups come in various shapes and sizes. Measure the rear recessed dimensions of the terminal cup and cut the correct size opening in the back of the panel.*

All holes are 3/8" in from edge of back panel.

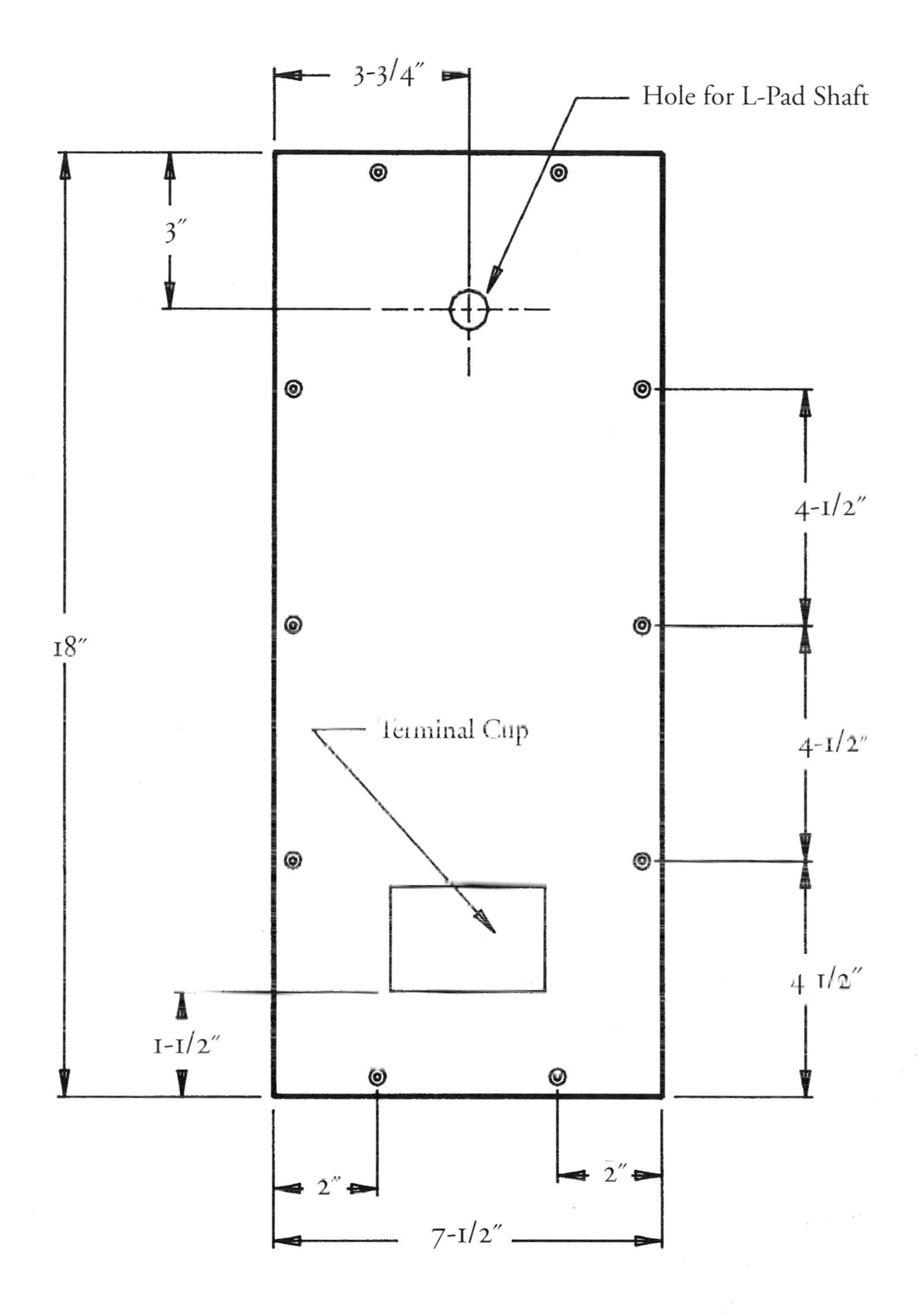
3-3/4″
Hole for L-Pad Shaft
3″
18″
4-1/2″
4-1/2″
Terminal Cup
4-1/2″
1-1/2″
2″
2″
7-1/2″

Back Panel

Material 3/4" MDF (medium-density fiberboard), particle board or plywood.

Size 3/4" x 7-1/2" x 13-3/4"

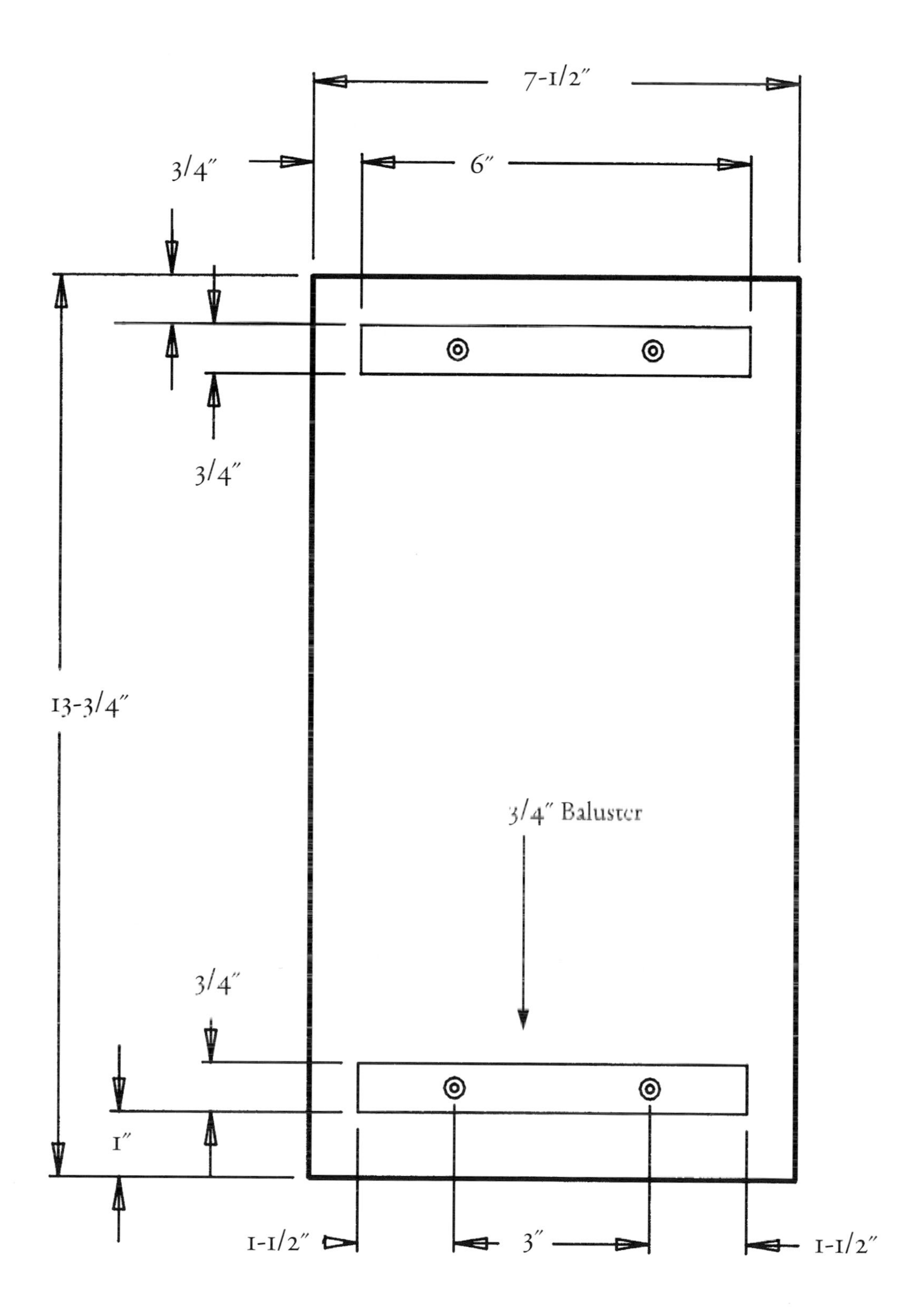

Top Panel (Interior View)

Material 3/4" MDF (medium-density fiberboard), particle board or plywood.

Size 3/4" x 7-1/2" x 13-3/4"

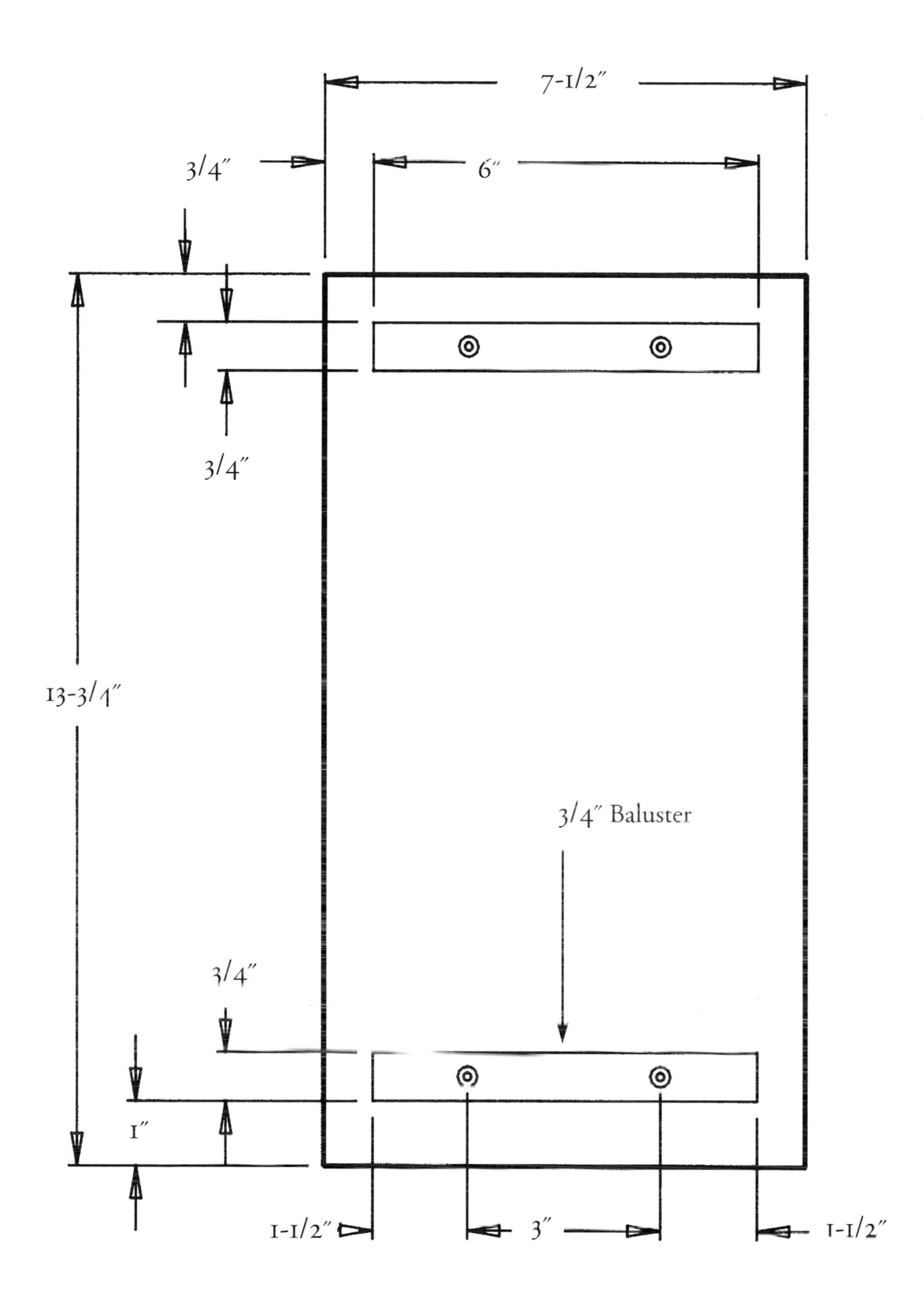

Bottom Panel (Interior View)

Material 3/4" MDF (medium-density fiberboard), particle board or plywood.

Size 3/4" x 13-3/4" x 19-1/2"

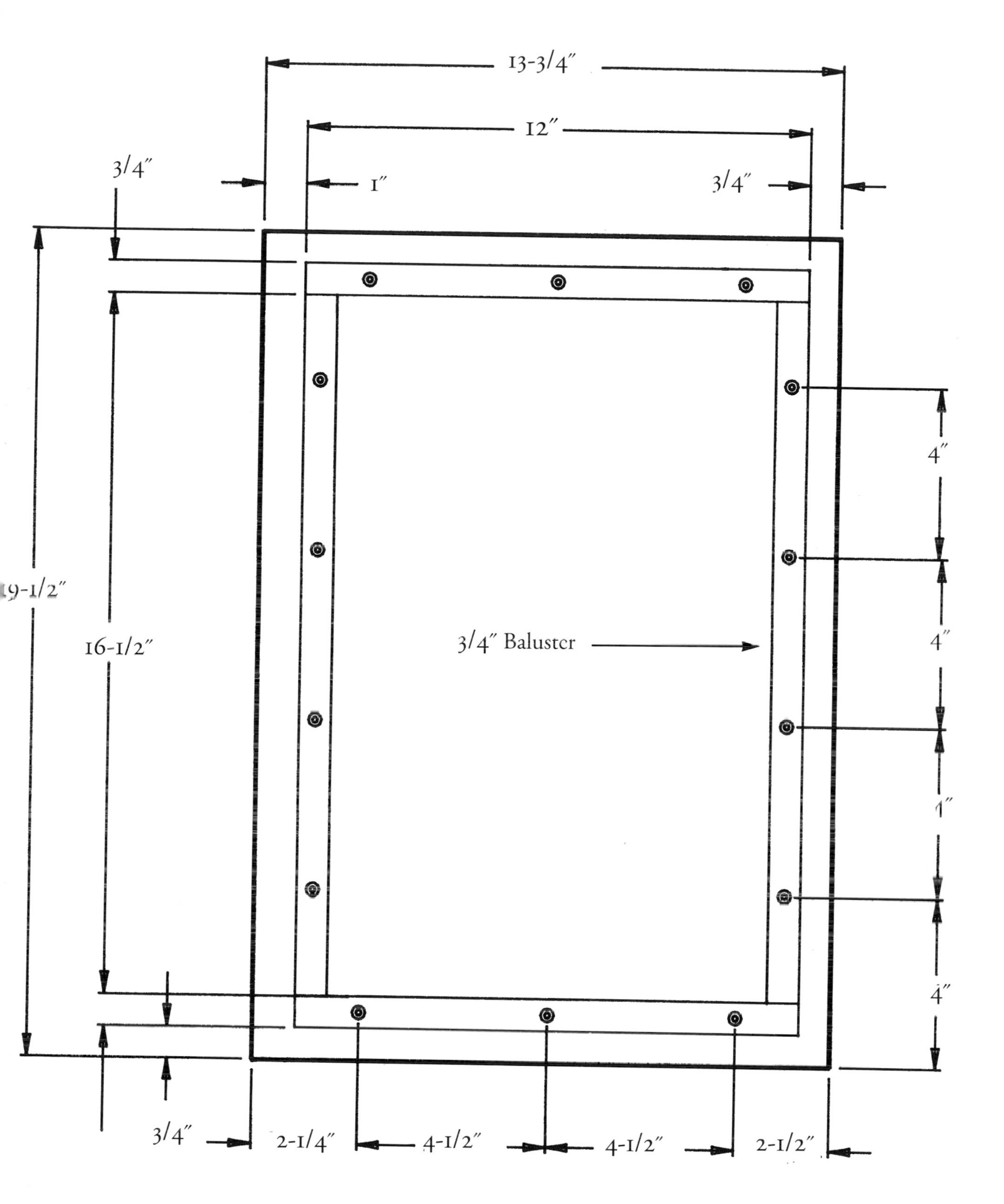

Left Side Panel (Interior View)

Material 3/4" MDF (medium-density fiberboard), particle board or plywood.

Size 3/4" x 13-3/4" x 19-1/2"

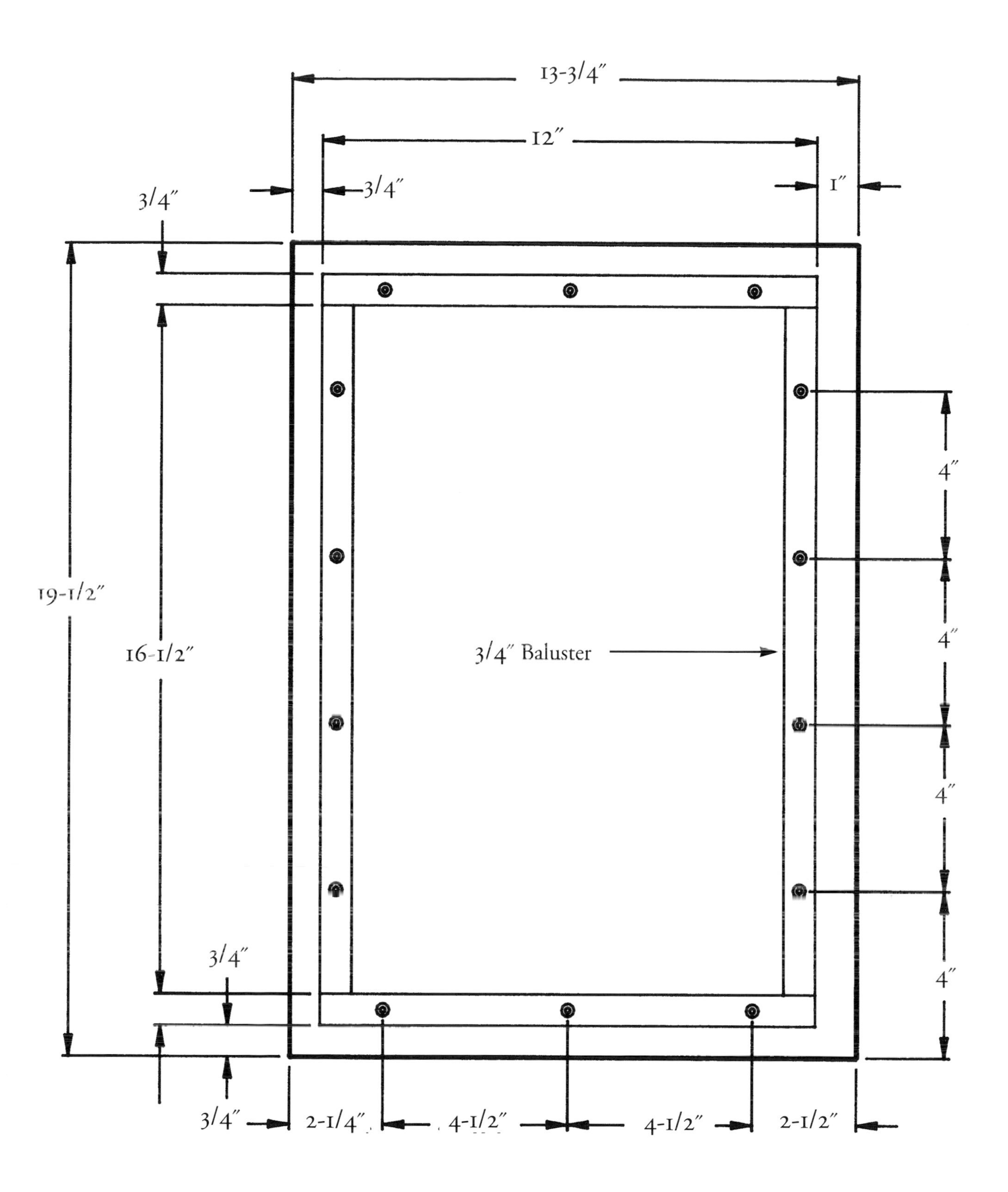

Right Side Panel (Interior View)

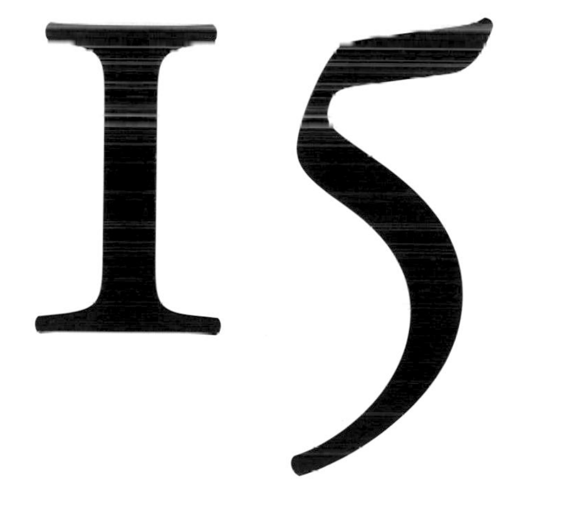

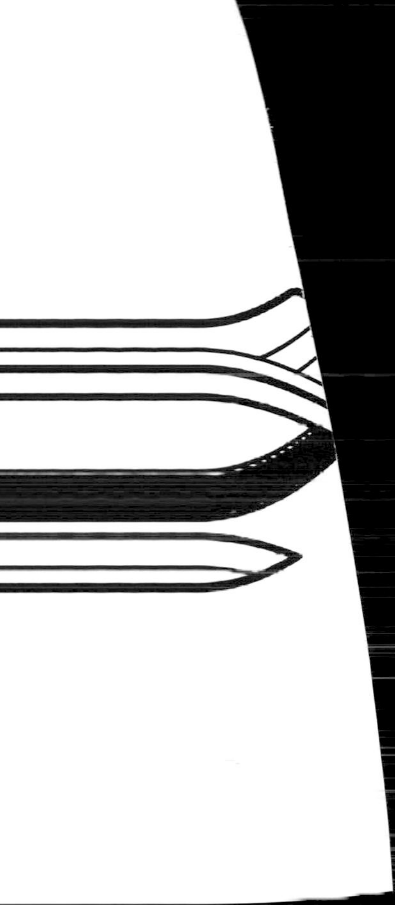

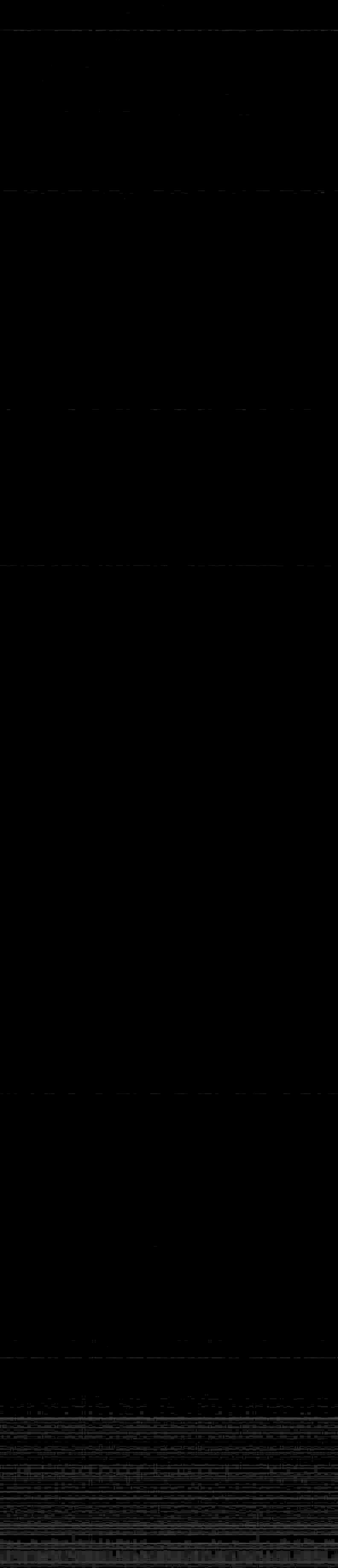

16

Installing the Components

Installing the speakers, port tube and electrical components into your enclosures is a simple procedure. Go over the following material before the final assembly to make sure everything is correct.

After all openings in the front baffle and rear panel have been cut out and painted (finished), fasten the painted front baffle with screws to the enclosure. Remember, all holes should be pre-drilled and the front baffle and rear panel screw holes counter sunk. Attach the terminal cup, crossover and L-Pad to the back panel. The crossover should be fastened to the interior side of the rear panel with screws between the L-Pad and the terminal cup. We suggest that you position the high or tweeter section of the crossover directly under the L-Pad for ease in wiring.

Note: *Use silicone adhesive on the inside of the L-Pad to keep it from turning when you adjust it. Wait until the silicone adhesive has set before you wire.*

Be careful not to screw down the recessed terminal cup too tightly. If too much force is used, the plastic edge of the cup may crack.

Wire the crossover and L-Pad according to the wiring diagrams and make sure all your positive (+) and negative (-) connections are hooked up correctly. Use enough wire so that the woofer and tweeter wires can hang out of the openings. If the wires are too short it may be difficult to manipulate the female connectors onto the speaker terminal posts. Do not fasten the rear panel to the enclosure yet.

Cut and staple polyester damping materials into the enclosure. Make sure staples are driven into the enclosure securely.

Place the speaker enclosure face up and put the port tube, woofer and tweeter into the openings. If the port tube is loose, use a drop or two of silicone adhesive to secure it. Use the correct diameter 3/4" Phillips roundhead screws for fastening the speakers to the front baffle of the enclosure. Mark with a pencil where the screws go into the baffle. Remove the speakers and pre-drill all screw holes. Place the speakers back into the openings in the baffle and screw them in. To avoid damaging the speakers be extremely careful not to slip with the screwdriver or screw bit on your variable speed drill. Screw the speakers into the baffles securely but not too tightly.

Note: *Some tweeters have plastic rings (frames). Be very careful not to screw these types of tweeters in so tightly that the plastic cracks.*

Remove the speakers from the front baffle and place the rear panel into the back of the enclosure. Do not screw it in yet. Guide the woofer wires and the tweeter wires through the correct openings in the front panel. Woofer wires through woofer opening. Tweeter wires through tweeter opening. Double check to make sure you have done this correctly. Screw in the rear panel with two screws only. You may have to remove it to make adjustments and changes.

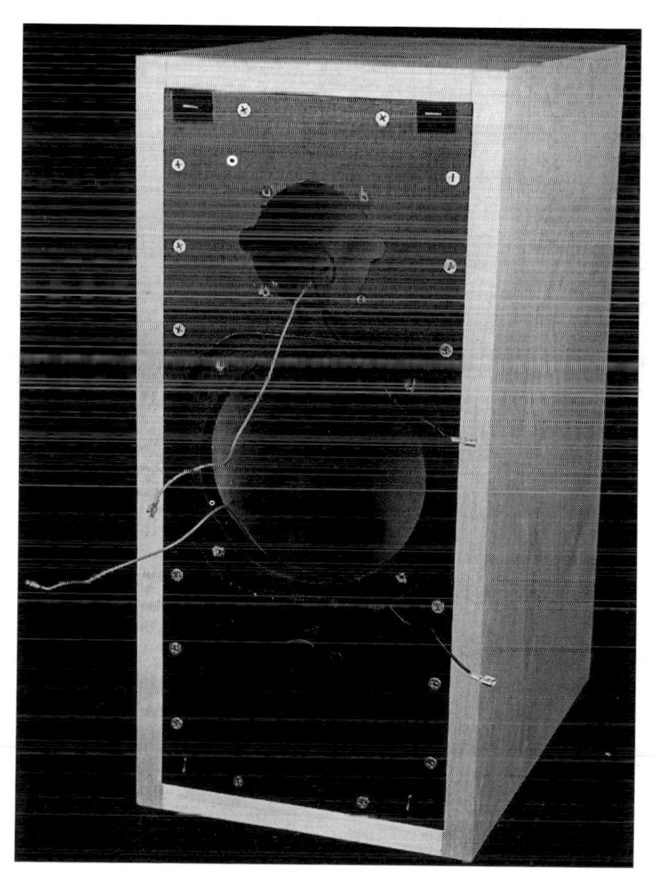

Enclosure with woofer and tweeter wires out openings.

Observe correct polarity when connecting speakers: positive (+) to positive (+) and negative (-) to negative (-).

Fasten the female connectors onto the speaker connecting posts and make sure to observe correct polarity. Positive speaker connecting posts are marked positive (+) and negative speaker connecting posts are marked negative (-). Sometimes the positive post is marked with a red color. Connections should be made with the red wire to positive (+) and the black wire to negative (-). After connecting the speakers, screw them into the front baffle carefully.

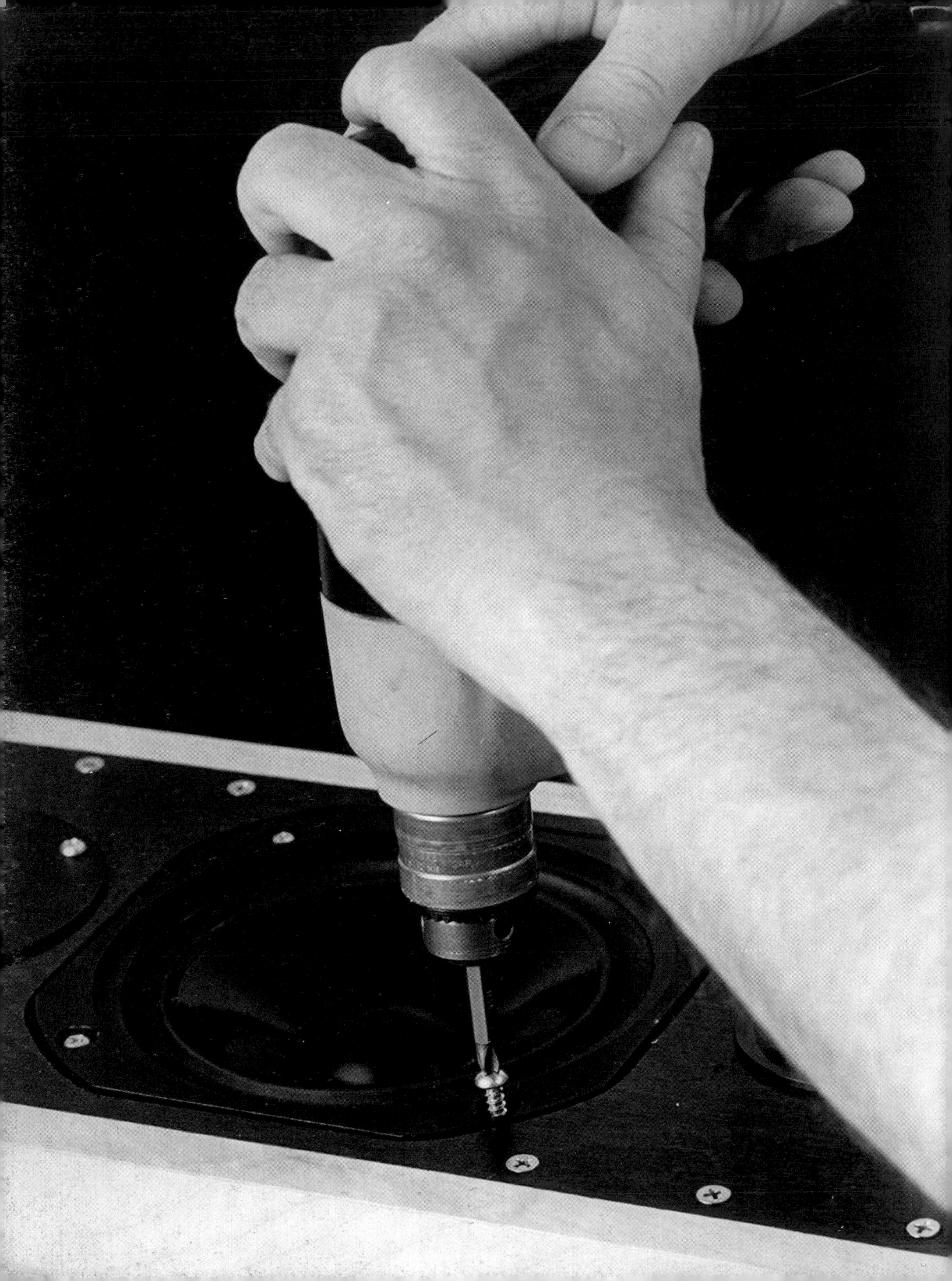

(Above) Touching up baffle screws with paint makes them less noticeable.
(Left) After connecting the speakers, screw them into the front baffle carefully.

17

The Speaker Grill

The speaker grill is easy to make. Cut out two pieces of 3/8" plywood to the outside dimensions of the front of your speaker enclosures (9" x 19-1/2"). Cut out the center of the plywood with a jigsaw following the measurements on the grill frame plan on page 139. Paint the cut-out grill frame black (or to match your front baffle) and stretch a measured piece of black or gray fiberglass window screen over the cut-out grill frame. Staple the edges of the screen onto the back of the painted grill frame. Attach the grill to your speaker baffle with pieces of adhesive-backed loop and hook strips and staples. Adhesive alone may not be strong enough to hold the grill in place. Staple loop and hook strips onto the baffle, not the enclosure frame around the baffle.

Fiberglass window screen is an excellent choice for speaker grills. It is acoustically transparent and will not affect the performance of loudspeakers. It also looks great. If you want your speakers to fit into the decor of your home, you can choose any color loosely woven, synthetic fabric. Stay away from natural fibers such as linen, wool and cotton. They shrink and can warp the grill frame.

Note: *You don't have to have a speaker grill. Some people like the look and sound of their speakers without grills.*

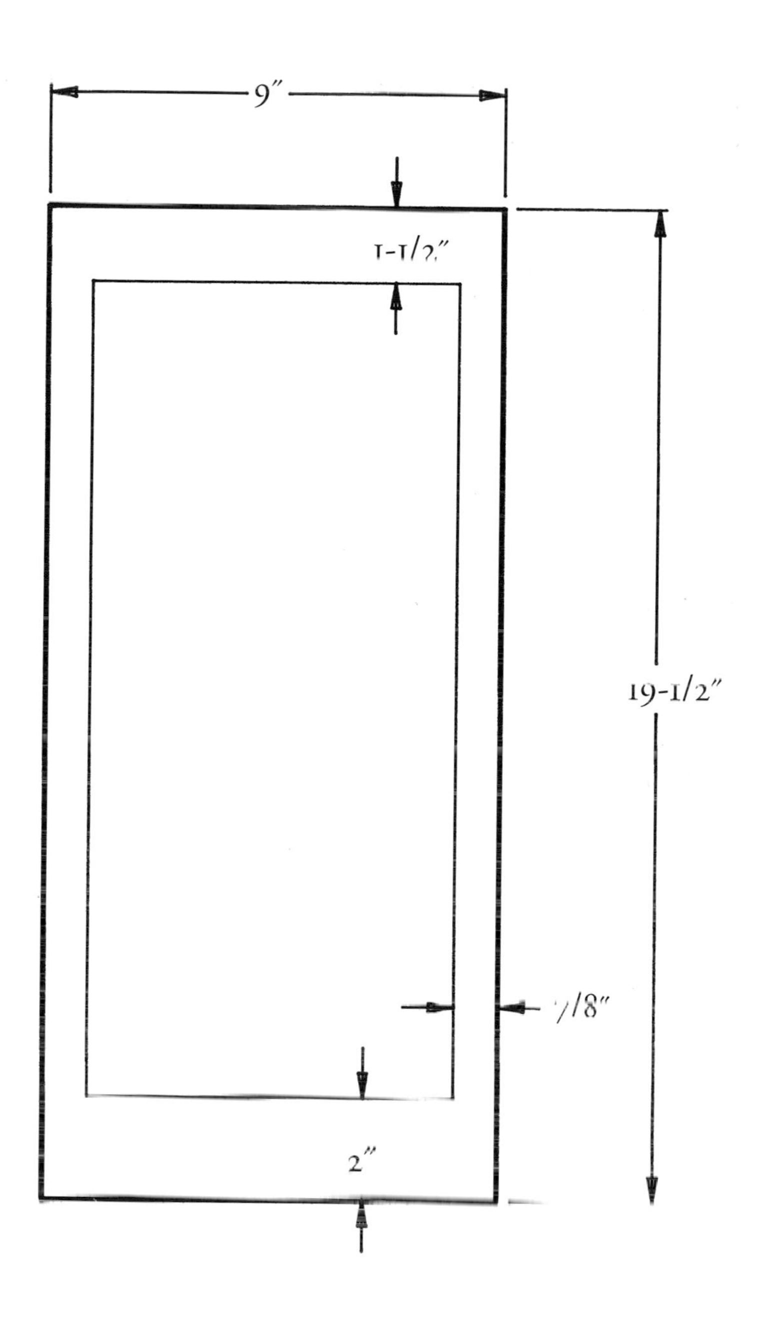

Grill Frame

Material Plywood

Size 3/8" x 9" x 19-1/2"

Stretch and staple fabric to the reverse side of the speaker grill.

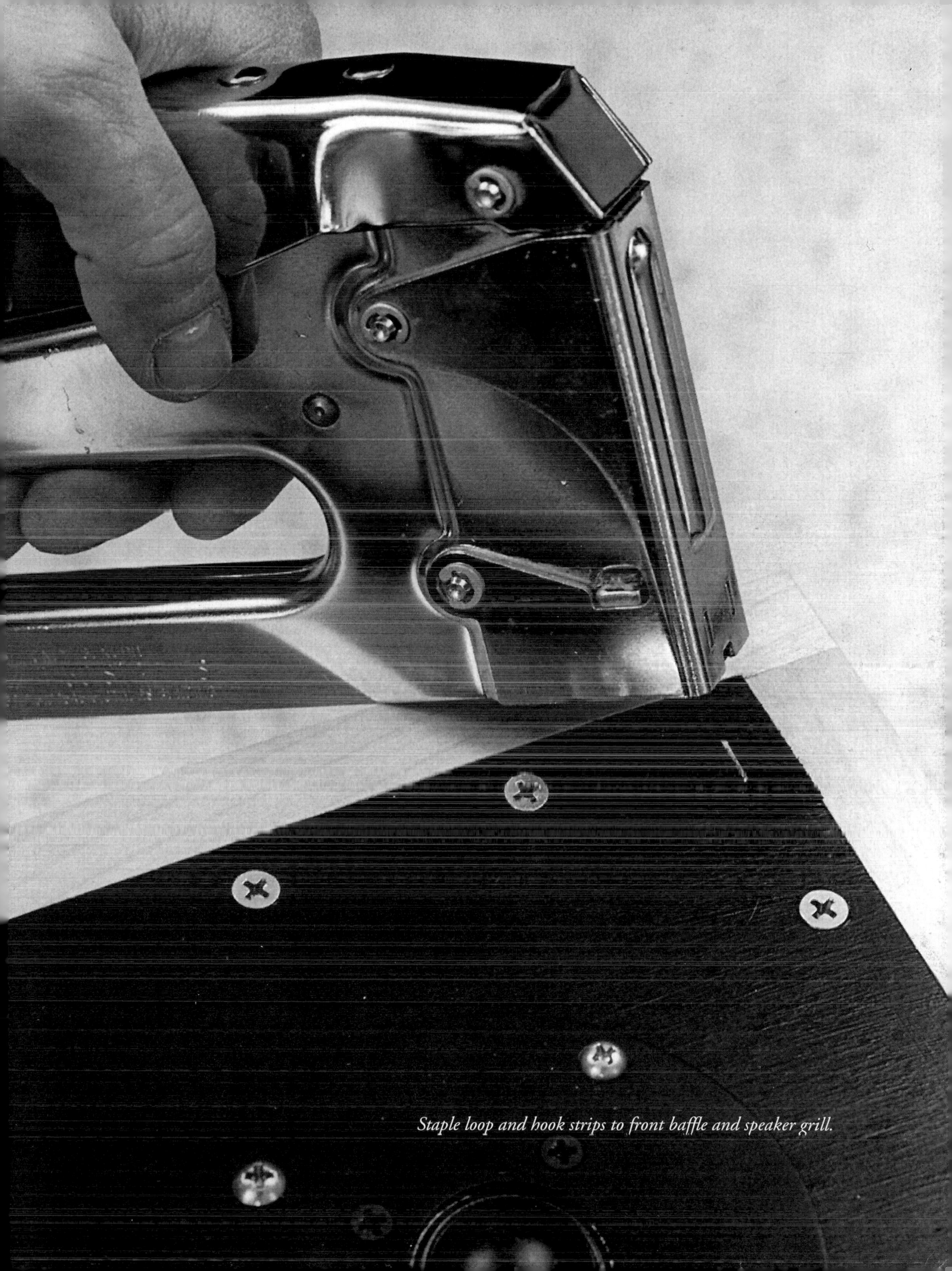

Staple loop and hook strips to front baffle and speaker grill.

Speaker grill stretched and stapled with loop and hook fasteners on the grill and front baffle.

18

Polarity Check

Checking the polarity of your woofers can be accomplished by touching two wires to the ends of a size D flashlight battery.

Attach a red wire to the positive (+) terminal of the terminal cup and a black wire to the negative (-) terminal of the terminal cup.

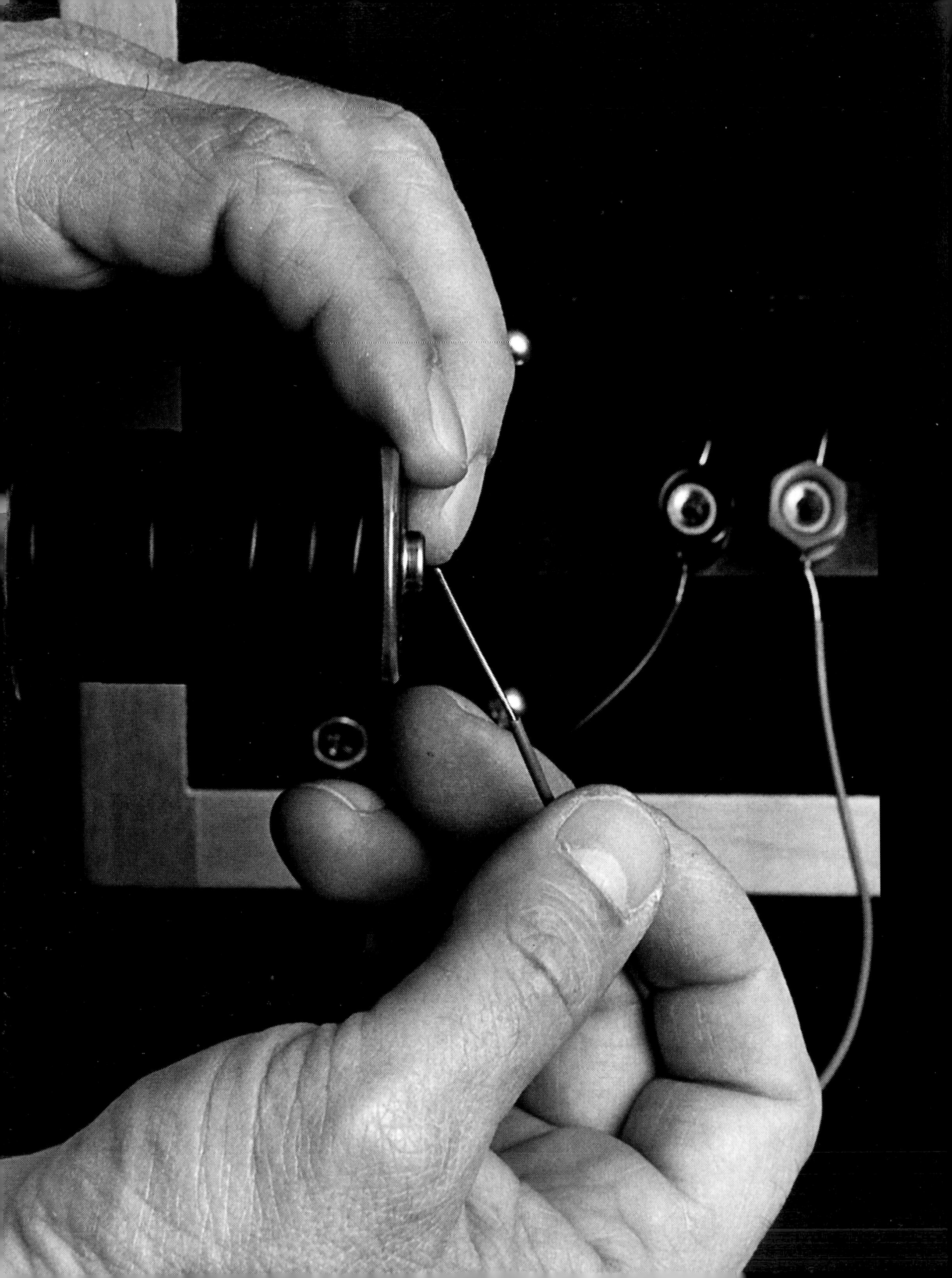

Touch the red wire to the positive (+) end of the battery and the black wire to the negative end of the battery. When you do this, look at the woofer. If the polarity is correct, the woofer will move forward. If the polarity is incorrect, the woofer will move backward. If the polarity is incorrect, simply change the connecting wires to the woofer and test again.

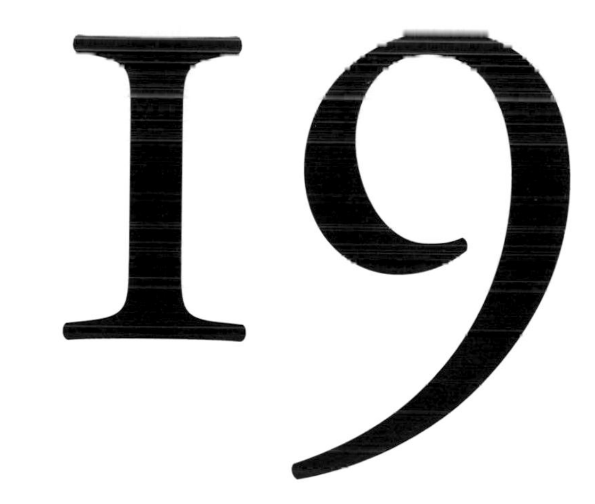

Loudspeaker Placement

Like all loudspeakers, all rooms have different acoustic (sound) characteristics. A heavily furnished room with wall-to-wall carpeting, drapes and padded furniture will not be as bright acoustically as a sparely furnished room with a bare hardwood floor and less padded furniture. Low sound frequencies (bass) will be more apparent in a heavily furnished room because the high sound frequencies will tend to be absorbed by the fabrics in the room. In a sparely furnished room, high sound frequencies will be emphasized because there are fewer furnishings to absorb them.

A speaker placed in a corner or against the wall of a room will produce more apparent bass than when it is moved away from the corner or wall. Placing your loudspeakers at ear level on a tightly packed bookshelf can accentuate bass response.

The traditional method of loudspeaker placement is to put the speakers on stands or secure platforms away from walls and room corners with the speakers facing toward thc listener and the tweeters at ear level. The distances between the speakers and the listener should be equal. However, in the real world, given the restraints of your environment, this may not be practical. It is also possible that you won't like the way the speakers sound when they are arranged in this manner.

Depending upon what kind of sound you like, it's up to you to determine where your speakers should be placed. The sound of your loudspeakers will change everytime you move them.

Note: *It takes time for a new woofer to break in. Bass response can improve by 10% and even more after 24 hours of playing time.*

20

Connecting Loudspeakers to an Amplifier or Receiver

Caution: *Make sure that you check the power capabilities of the speakers in your system and make sure that you do not drive the speakers beyond the power limits set by the manufacturer!*

Before connecting your speakers to an amplifier or receiver make sure the power is turned off and the volume control is turned all the way down.

Use commercially sold loudspeaker connecting wire for hooking up your amplifier or receiver to your loudspeakers. Sixteen gauge wire is good—18 gauge will do for shorter lengths. Strip the insulation off the ends of the wires before making connections. If you purchase wire that is marked or color- coded for positive and negative connections, it will be easier to connect your speakers correctly.

The back of your receiver will have terminal connectors marked for connecting each speaker. The terminals will say "Right" and "Left." One set of connectors will be for the right loudspeaker and one set for the left. Connect the positive (+) wire lead to the positive (+) connector (usually colored red on the back of the amplifier or receiver) to the positive (+) (red) terminal on the terminal cup. Hook the other wire lead (the negative (-) terminal usually colored black on the back of your amplifier or receiver) to the negative (-) (black) terminal on your terminal cup. Follow this procedure for both speakers. If connections are made improperly, the loudspeakers will be out of phase and performance will suffer.

Note: *As a rule, always turn the volume level of your amplifier or receiver down after you finish listening to loudspeakers. One of the most common causes of loudspeaker damage is sudden bursts of power caused by elevated volume levels.*

If you are concerned about damaging your speakers at high power levels, you can purchase loudspeaker fuse holders and fuses. Use "Fast Acting Fuses." Fuse holders and fuses can be connected to the positive (+) (red) terminal on the terminal cup and the end of the positive (+) lead wire coming from the amplifier or receiver. To determine the proper fuse, refer to the following chart. The fuse ratings shown are conservative in order to provide an extra margin of safety.

Speaker Power Capacity	Fuse size (amp)
20 watts	1 amp
30 watts	1-1/2 amp
40 watts	2 amp
50 watts	2 amp
60 watts	2-1/2 amp
70 watts	2-1/2 amp
80 watts	3 amp
90 watts	3 amp
100 watts	3 amp

Note: *The use of fuses may cause a slight (probably imperceptible) change in the way your speakers sound. If you won't be playing your loudspeakers at high volume or power levels, you won't need fuses.*

21

Where to Buy Speakers and Components (Ask for catalogues)

A&S Speakers
4075 Sprig Drive
Concord, CA 94520
Phone: 510-685-5252

Madisound Speaker Components
P.O. Box 44283
Madison, WI 53744
Phone: 608-831-3433
FAX: 608-831-3771

McBride Loudspeaker Source
638 Colby Drive
Waterloo, ONT N2V 1A2
Canada
Phone: 519-884-3500

MCM Electronics
650 Congress Park Drive
Centerville, Ohio 45459
Phone: 800-543-4330
FAX: 513-434-6959

Meniscus
2575 28th Street S.W., #2
Wyoming, MI 49509
Phone: 616-543-9121
FAX: 616-534-7676

Parts Express
340 East First Street
Dayton, Ohio 45402
Phone: 1-800-338-0531
FAX: 513-222-4644

Solen, Inc.
4470 Thibault Avenue
St-Hubert, QC J3Y 7T9
Canada
Phone: 514-656-2759
FAX: 514-443-4949

Speaker City
115 South Victory Blvd.
Burbank, CA 91502
Phone: 818-846-9921

Speakers, Etc.
2730 West Thomas Road
Phoenix, AZ 85017
Phone: 602-272-6696
FAX: 602-272-8718

Zalytron Industries Corp.
469 Jericho Turnpike
Mineola, NY 11501
Phone: 516-747-3515
FAX: 516-294-1943

Order Information

THE AUDIOPHILE LOUDSPEAKER *Anyone Can Build*

makes a great gift or stocking stuffer!

To purchase additional books send a check for $22.95 plus $3.50 shipping and handling ($4.95 in Canada) to:

Boston Post Publishing Co.
P.O. Box 1175
Madison, CT 06443

Or call: 1-800-507-2665
Have your Master/Visa Card ready.

If you wish to purchase more than one book, add $1.00 shipping and handling for each additional book.

Sorry, no C.O.D.
Please add 6% sales tax for books shipped to Connecticut addresses.

Stay informed. Send your name and address and we'll send you literature on all upcoming projects free of charge.